A NOVEL APPROACH TO SLUDGE TREATMENT USING MICROWAVE TECHNOLOGY

Eva Kocbek

A NOVEL APPROACH TO SLUDGE TREATMENT USING MICROWAVE TECHNOLOGY

DISSERTATION

Submitted in fulfillment of the requirements of
the Board for Doctorates of Delft University of Technology
and
of the Academic Board of the IHE Delft
Institute for Water Education
for
the Degree of DOCTOR
to be defended in public on
Friday, 5 November 2021, at 10.00 hours
in Delft, the Netherlands

by

Eva KOCBEK
Master of Science in Urban Water and Sanitation; Specialization Water Supply Engineering,
UNESCO-IHE Institute for Water Education, Delft, The Netherlands
born in Maribor, Slovenia

This dissertation has been approved by the
promotor: Prof. dr. D. Brdjanovic and
copromotor: Dr. H.A. Garcia Hernandez

Composition of the doctoral committee:

Rector Magnificus TU Delft	Chairman
Rector IHE Delft	Vice-Chairman
Prof.dr. D. Brdjanovic	IHE Delft / TU Delft, promotor
Dr. H.A. Garcia Hernandez	IHE Delft, copromotor

Independent members:
Prof.dr. A. Léonard	University of Liège, Belgium
Prof.dr. T. Koottatep	Asian Institute of Technology, Thailand
Prof.dr.sc. M. Matosic	University of Zagreb, Croatia
Prof.dr.ir. M.K. de Kreuk	TU Delft
Prof. dr.ir. J.B. van Lier	TU Delft, reserve member

This research was conducted under the auspices of the Graduate School for Socio-Economic and Natural Sciences of the Environment (SENSE)

CRC Press/Balkema is an imprint of the Taylor & Francis Group, an informa business

Published by:
CRC Press/Balkema
enquiries@taylorandfrancis.com
www.crcpress.com – www.taylorandfrancis.com
ISBN 978-1-032-21799-4.

To

my father

Summary

As most cities are engaging in an ongoing rapid process of urbanization, and the corresponding increase in population, waste generation, and urban settlement, multidimensional issues have emerged concerning the management of municipal sewage and faecal sludge. These problems have resulted in a series of health concerns as ineffective management of these forms of sludge can result in pathogens and contaminants entering terrestrial ecosystems, leading to the contamination of water resources and disease outbreaks, among other issues. The situation is particularly complicated when large amounts of sludge that contain a variety of contaminants are produced in areas characterized with a rapid accumulation of sludge, such as emergency settlements, slums and other densely populated settlements. The establishment of fast and effective sludge management approaches are essential in these settings.

In more recent times, researchers have examined an array of technological means of recovering fertilizer, energy, and water from on-site sanitation and wastewater treatment facilities while aiming for pathogen inactivation and/or removal. A group of researchers (Mawioo et al., 2016a; 2016b; 2017) suggested a novel method of treating sludge that could make it possible to inactivate and/or destroy the pathogens and in the same time increase the energetic potential of sludge from sewage treatment, fresh faecal sludge and septic tank sludge by the application of microwave technology.

A limited number of lab-based studies so far have found that microwave energy is a fast and efficient method for sludge treatment, which can facilitate sterilization of sludge and reduce sludge mass and volume by up to 5% over a shorter time duration than the conventional sludge drying technologies. The prospective benefits of microwave treatment have prompted scientists to invest efforts in studying this promising technology and amid efforts to design and establish the microwave system's performance in sludge treatment at a pilot-scale level. However, when attempting to scale-up a microwave system from lab-to pilot-scale, the system's energy performance in the treatment of sludge was low, and considerable operational cost in terms of specific energy consumption (SEC) were reported during the drying process (> 16 MJ L^{-1} or 4 kWh L^{-1}).

The first objective of this research is to develop and implement solutions that can adequately address the existing issues associated with the microwave technology observed in the previous studies. The solutions include measures that can relatively easily extract vapour generated during the use of the system and augment electromagnetic distribution uniformity within the sample. The novel microwave pilot-scale system's performance was evaluated in the treatment of mechanically (centrifugation) dewatered sludge and assessed through monitoring the achieved throughput capacity and specific energy consumed during the drying process using microwaves. The results showed that the technical solutions introduced had a beneficial impact on the system's energy performance; a 70% decrease in the specific energy consumption was achieved compared to microwave sludge drying studies elsewhere (from 16 MJ L^{-1} or 4.4 kWh L^{-1} to 4.3 MJ L^{-1} or

1.2 kWh L^{-1}). The results showed that the specific energy consumption was dependent on the conversion efficiency of the system from electric energy to electromagnetic energy (i.e., microwave generation efficiency), which varied because of changes in the microwave output power. The lowest specific energy consumption observed was obtained at the maximum microwave output power, generating the highest microwave generation efficiency. The study outcomes indicated that industrial microwave generators with a high microwave conversion efficiency should be implemented to obtain optimal results, and should be operated at maximum capacity. The results also indicated that it might be possible to achieve a further reduction in specific energy consumption by recovering energy from the condensate that is generated during the process of drying to subsequently preheat the incoming sludge.

Another fundamental aspect that influenced the electrical demand of the microwave system and throughput capability is the power density, i.e., the amount of microwave energy absorbed by sample mass or volume. The microwave unit's performance was enhanced by adjusting the power-to-mass ratio to adapt the power density across the sample. More precisely, the utilization of high microwave power outputs and a reduced applied mass during treatment decreased the exposure time, increased the rate of drying and reduced the amount of energy consumed for water removal. The results also indicated that the application of a low sample mass in combination with high powers might have also led to enhanced reflective power energy losses, thereby undermining the energy system's performance. This situation worsened when the depth of the sludge layer exposed to the radiation was considerably higher than the level to which the microwaves can penetrate. The thicker the sample, the greater the extent of the lack of uniformity of the electromagnetic field that is distributed throughout the sample. This subsequently led to a reduction in the process energy efficiency with a corresponding increase in specific energy consumption. The non-uniformity of electromagnetic energy strength along the material depth was, however, to a certain extent mitigated by the effect of moisture levelling, which is the preferential absorption of microwaves in material regions with higher moisture content. In other words, while the uniformity and efficiency of the microwave energy that is absorbed by the material over a given period represent a vital aspect in increasing the energy performance of the system, it is the characteristics of the material processed, specifically the dielectric properties of sludge, that regulate and determine the absorption of the microwave. This research demonstrated that water molecules' quantity and distribution are among the main parameters that govern the sludge dielectric properties. Specifically, the findings demonstrated that the water not attached to the sludge matrix moved (rotated) more efficiently under the applied microwave field than bound water molecules, causing enhanced water removal from the material. The research also established that the distribution of free and bound water was, to a certain extent, dependent on the sludge hydrophobicity as it is influenced by its fat and oil content. The selected sludge samples containing higher fat and oil content were more hydrophobic and thus held less available sites for absorption by bound water. The latter also proved to have a positive impact on the energetic content of the sludge.

Overall, these findings have advanced a fundamental understanding of the relationship between microwave energy and sludge and have shown that the advantages consistently cited in the literature, such as volumetric and selective heating, effectively result in rapid processing of sludge with a relatively low physical footprint requirement and competitive energy use in comparison to other thermal drying technologies.

The system's high throughput capacity at low required contact/surface area for heat and mass transfer between the sludge samples and the microwave energy enabled the system to be containerized (manufactured in a single container) and mounted on the trailer to perform *in-situ* sludge treatment. The potential benefits of using microwave technology in the *in-situ* treatment of sludge from on-site sanitation facilities have been successfully tested and confirmed in the field under a real scenario in Jordan Valley (Jordan). Results obtained from the field testing demonstrated that the microwave technology effectively reduced the level of pathogens and moisture in sludge obtained from septic tanks, and thus positively impacted the potential use of sludge, which depending on the composition, may be directly used for soil amendment or fuel. Therefore, the key shortfall leading to a large bottleneck (i.e., the sludge moisture content) hindering the sludge management efficiency has been addressed through the application of microwave heating. Besides microwave technology, several other technological solutions have been implemented into the mobilized treatment system, including; mechanical dewatering technique assisted by conditioning agents and ultrafiltration units followed by reverse osmosis. The innovative use of technology incorporated in the mobile system has made it possible to reduce the cost of microwave sludge drying while at the same time allowing for the recovery of additional resources, including water and dried sludge residue that may be reused either for irrigation or industrial processes.

This doctoral thesis demonstrates that treatment of sludge using microwave technology in standalone treatment and with or without alternative solutions for liquid and solid treatment is a sustainable technological option for sludge derived from wastewater treatment and on-site sanitation, generating products that may be reused.

References

Mawioo PM, Garcia HA, Hooijmans CM, Velkushanova K, Simonič M, Mijatović I, Brdjanovic D. (2017). A pilot-scale microwave technology for sludge sanitization and drying. Science of the Total Environment, 601, 1437-1448.

Mawioo PM, Hooijmans CM, Garcia HA, Brdjanovic D. (2016a). Microwave treatment of faecal sludge from intensively used toilets in the slums of Nairobi, Kenya. Journal of Environmental Management, 184, 575-584.

Mawioo PM, Rweyemamu A, Garcia HA, Hooijmans CM, Brdjanovic D. (2016b). Evaluation of a microwave based reactor for the treatment of blackwater sludge. Science of the Total Environment, 548, 72-81.

Samenvatting

Aangezien de meeste steden zich bezighouden met een voortdurend snel verstedelijkingsproces, en de overeenkomstige toename van de bevolking, afvalproductie en stedelijke nederzettingen, zijn er multidimensionale problemen ontstaan met betrekking tot het beheer van gemeentelijk rioolwater en fecaal slib. Deze problemen hebben geleid tot een reeks gezondheidsproblemen, aangezien ondoelmatig beheer van deze vormen van slib ertoe kan leiden dat ziekteverwekkers en verontreinigende stoffen terrestrische ecosystemen binnendringen. Dit kan onder meer leiden tot de vervuiling van watervoorraden en het uitbreken van ziekten. De situatie is bijzonder gecompliceerd wanneer grote hoeveelheden slib met een verscheidenheid aan verontreinigingen worden geproduceerd in gebieden die worden gekenmerkt door een snelle ophoping van slib, zoals noodnederzettingen, sloppenwijken en andere dichtbevolkte nederzettingen. Het opzetten van snelle en effectieve slibbeheerbenaderingen is essentieel in deze omgevingen.

Recentelijk hebben onderzoekers een reeks technologische middelen onderzocht om kunstmest, energie en water terug te winnen uit lokale sanitaire voorzieningen en afvalwaterzuiveringsinstallaties, waarbij werd gestreefd naar inactivering en/of verwijdering van pathogenen. Een groep onderzoekers researchers (Mawioo et al., 2016a; 2016b; 2017) stelde een nieuwe methode voor de behandeling van slib die het mogelijk zou maken om de ziekteverwekkers te inactief te maken en/of te vernietigen en tegelijkertijd het energetisch potentieel van slib uit rioolwaterzuivering, fecaal slib en septageslib door de toepassing van microgolftechnologie te verhogen.

Een beperkt aantal laboratoriumstudies heeft tot dusverre aangetoond dat microgolfenergie een snelle en efficiënte methode is voor slibbehandeling, die sterilisatie van slib kan vergemakkelijken en ook slibmassa en -volume tot 5% kan verminderen in een kortere tijdsduur dan de conventionele technologieën voor het drogen van slib. De toekomstige voordelen van microgolfbehandeling hebben wetenschappers ertoe aangezet om te investeren in het bestuderen van deze veelbelovende technologie en hierbij de prestaties van het microgolfsysteem bij slibbehandeling op pilotschaal te ontwerpen en vast te stellen. Bij een poging om een microgolfsysteem op te schalen van laboratorium- naar pilootschaal, waren de energieprestaties van het systeem bij de behandeling van slib laag en werden aanzienlijke operationele kosten in termen van specifiek energieverbruik (SEC) gerapporteerd tijdens het droogproces. (> 16 MJ L^{-1} or 4 kWh L^{-1}).

Het eerste doel van dit onderzoek is het ontwikkelen en implementeren van oplossingen die de bestaande problemen in verband met de microgolftechnologie die in de vorige studies zijn waargenomen, adequaat kunnen aanpakken. De oplossingen omvatten maatregelen die relatief gemakkelijk damp kunnen extraheren die tijdens het gebruik van het systeem wordt gegenereerd, en die de uniformiteit van de elektromagnetische distributie binnen het monster vergroten. De prestaties van het nieuwe microgolf-pilootschaalsysteem werden geëvalueerd bij de behandeling van mechanisch (centrifugatie) ontwaterd slib. Deze werd beoordeeld door het monitoren van de bereikte doorvoercapaciteit en het specifieke energieverbruik tijdens het droogproces met behulp van microgolven. De resultaten toonden aan dat de geïntroduceerde technische oplossingen een gunstige invloed hadden op de energieprestaties van het systeem; een daling van 70% in het specifieke energieverbruik werd bereikt in vergelijking met microgolf-slibdroogonderzoeken elders (van 16 MJ L^{-1} or 4.4 kWh L^{-1} to 4.3 MJ L^{-1} or 1.2 kWh L^{-1}). Ook toonden de resultaten aan dat het specifieke energieverbruik afhankelijk was van het omzettingsrendement van het systeem van elektrische energie naar elektromagnetische energie (d.w.z. efficiëntie van microgolfopwekking) die varieerde vanwege veranderingen in het microgolfuitgangsvermogen. Het laagste specifieke energieverbruik dat werd waargenomen, werd verkregen bij het maximale microgolfuitgangsvermogen, waardoor de hoogste efficiëntie van microgolfopwekking werd gegenereerd. De onderzoeksresultaten gaven aan dat industriële microgolfgeneratoren met een hoge microgolfconversie-efficiëntie moeten worden geïmplementeerd om optimale resultaten te verkrijgen en op maximale capaciteit moeten worden gebruikt. De resultaten gaven ook aan dat het mogelijk zou zijn om een verdere verlaging van het specifieke energieverbruik te bereiken door energie terug te winnen uit het condensaat dat tijdens het drogen wordt gegenereerd om vervolgens het binnenkomende slib voor te verwarmen.

Een ander fundamenteel aspect dat de elektrische vraag van het microgolfsysteem en de doorvoercapaciteit beïnvloedde, is de vermogensdichtheid, d.w.z. de hoeveelheid microgolfenergie die wordt geabsorbeerd door de massa of het volume van het monster. De prestaties van de microgolfunit werden verbeterd door de vermogen-massaverhouding aan te passen om de vermogensdichtheid over het monster aan te passen. Preciezer, het gebruik van een hoog microgolfvermogen en een verminderde toegepaste massa tijdens de behandeling verminderde de belichtingstijd, verhoogde de droogsnelheid en verminderde de hoeveelheid energie die werd verbruikt voor het verwijderen van water. De resultaten gaven ook aan dat de toepassing van een lage monstermassa in combinatie met hoge vermogens ook zou kunnen hebben geleid tot grotere energieverliezen door reflecterend vermogen, waardoor de prestaties van het energiesysteem zijn ondermijnd. Deze situatie verslechterde toen de diepte van de sliblaag die aan de straling werd blootgesteld aanzienlijk hoger was dan het niveau waartoe de microgolven kunnen doordringen. Hoe dikker het monster, hoe groter het gebrek aan uniformiteit van het elektromagnetische veld dat door het monster wordt verdeeld. Dit leidde vervolgens tot een vermindering van de energie-efficiëntie van het proces met een overeenkomstige toename van het specifieke energieverbruik. De niet-uniformiteit van de elektromagnetische energiesterkte langs de materiaaldiepte werd echter tot op zekere hoogte verzacht door het effect van vochtnivellering. Dit is de voorkeursabsorptie van microgolven in materiaalgebieden met een hoger vochtgehalte. Met andere woorden, hoewel de uniformiteit en efficiëntie van de microgolfenergie die gedurende een bepaalde periode door het materiaal wordt geabsorbeerd een essentieel aspect vormen bij het verbeteren van de energieprestaties van het systeem, zijn het de kenmerken van het verwerkte materiaal, met name de diëlektrische eigenschappen van slib, dat de opname van de microgolf reguleert en bepaalt. Dit onderzoek toonde aan dat de hoeveelheid en distributie van watermoleculen de belangrijkste parameters zijn die de diëlektrische eigenschappen van slib bepalen. Specifiek toonden de bevindingen aan dat het water dat niet aan de slibmatrix was gehecht, efficiënter onder het toegepaste microgolfveld bewoog (geroteerd) dan gebonden watermoleculen, waardoor het water beter uit het materiaal werd verwijderd. Het onderzoek toonde ook aan dat de verdeling van vrij en gebonden water tot op zekere hoogte afhankelijk was van de hydrofobiciteit van het slib, aangezien dit wordt beïnvloed door het vet- en oliegehalte. De geselecteerde slibmonsters met een hoger vet- en oliegehalte waren meer hydrofoob en hielden dus minder beschikbare plaatsen vast voor opname door gebonden water. Dit laatste bleek ook een positieve invloed te hebben op het energetisch gehalte van het slib.

Al met al hebben deze bevindingen geleid tot een fundamenteel begrip van de relatie tussen microgolfenergie en slib en hebben ze aangetoond dat de voordelen die consequent in de literatuur worden genoemd, zoals volumetrische en selectieve verwarming, effectief resulteren in een snelle verwerking van slib met een relatief lage fysieke voetafdruk en concurrerend energieverbruik in vergelijking met andere thermische droogtechnologieën.

Door de hoge doorvoercapaciteit van het systeem bij een laag vereist contact/oppervlak voor warmte- en massaoverdracht tussen de slibmonsters en de microgolfenergie kon het systeem worden gecontaineriseerd (vervaardigd in een enkele container) en op de trailer gemonteerd om *in-situ* slibbehandeling uit te voeren. De potentiële voordelen van het gebruik van microgolftechnologie bij de *in-situ* behandeling van slib uit lokale sanitaire voorzieningen zijn met succes getest en bevestigd in het veld onder een reëel scenario in Jordan Valley (Jordanië). Resultaten verkregen uit de veldtesten toonden aan dat de microgolftechnologie het niveau van ziekteverwekkers en vocht in slib uit septic tanks effectief verlaagde en dus een positief effect had op het potentiële gebruik van slib, dat, afhankelijk van de samenstelling, direct kan worden gebruikt voor bodemverbetering of brandstof. Daarom is het belangrijkste tekort dat leidt tot een groot knelpunt (d.w.z. het vochtgehalte van het slib) dat de efficiëntie van slibbeheer belemmert, aangepakt door de toepassing van microgolfverwarming. Naast microgolftechnologie zijn er verschillende andere technologische oplossingen geïmplementeerd in het gemobiliseerde behandelingssysteem, waaronder; mechanische ontwateringstechnieken ondersteund door conditioneringsmiddelen en ultrafiltratie-eenheden gevolgd door omgekeerde osmose. Het innovatieve gebruik van technologie die in het mobiele systeem is ingebouwd, heeft het mogelijk gemaakt om de kosten van het drogen van slib in de magnetron te verlagen en tegelijkertijd de terugwinning van extra hulpbronnen mogelijk te maken, waaronder water en gedroogd slibresten die kunnen worden hergebruikt voor irrigatie of industriële processen.

Dit proefschrift toont aan dat de behandeling van slib met behulp van microgolftechnologie in een op zichzelf staande behandeling en met of zonder alternatieve oplossingen voor de behandeling van vloeistoffen en vaste stoffen een duurzame technologische optie is voor slib afkomstig van afvalwaterzuivering en lokale sanitatie, waarbij producten worden gegenereerd die kunnen worden hergebruikt

References

Mawioo PM, Garcia HA, Hooijmans CM, Velkushanova K, Simonič M, Mijatović I, Brdjanovic D. (2017). A pilot-scale microwave technology for sludge sanitization and drying. Science of the Total Environment, 601, 1437-1448.

Mawioo PM, Hooijmans CM, Garcia HA, Brdjanovic D. (2016a). Microwave treatment of faecal sludge from intensively used toilets in the slums of Nairobi, Kenya. Journal of Environmental Management, 184, 575-584.

Mawioo PM, Rweyemamu A, Garcia HA, Hooijmans CM, Brdjanovic D. (2016b). Evaluation of a microwave based reactor for the treatment of blackwater sludge. Science of the Total Environment, 548, 72-81.

Table of content

1

Introduction

1.1 General introduction

The pressure on the environment in the pursuit for resources and the lack of drinking water access, exacerbated by poorly managed wastewater treatment and disposal, pose some of the main challenges modern society is in urgent need to address (Corcoran et al., 2010). According to the World Health Organisation, almost 1.7 million people die every year from diarrhoeal diseases worldwide and more than a quarter of diarrheal disease is experienced by children below the age of five (WHO, 2017). One of the factors contributing to these alarming figures is the exponential growth rate of the World's population, which is projected to increase to approximately 9.7 billion by the end of 2050 (UN, 2019). The population projections show that the highest growth is expected to occur in urban areas, from 4.2 to 6.6 billion people (UN, 2018). The uncontrolled discharges of untreated faecal sludge and wastewater not only have an impact on the mortality and morbidity rates but also contribute towards perpetuating poverty and social inequalities through cost for medical care and hampered workforce productivity (Corcoran et al., 2010; Afolabi et al., 2017). For instance, the estimated annual costs spent on health care resulting from inadequate sanitation alone may be as high as USD 223 billion (Hutton et al., 2007). Whereas, a dollar invested in water and sanitation improvements, may result in savings of USD 5–46 depending on the intervention (Hutton et al., 2007).

Recognising the financial, environmental and societal costs of the impact of uncontrolled sludge and wastewater discharges on human health, international attention has shifted to increasing access to sanitation with the commitment expressed at the Millennium Summit of the United Nations in 2000, following the adoption of the United Nations Millennium Declaration that resulted in the Millennium Development Goals (MDGs) (Corcoran, 2010). MDGs' partial success as far as sanitation is concerned urged that Clean Water and Sanitation becomes one of the Sustainable Millennium Goals (Goal nr 6) to "Ensure availability and sustainable management of water and sanitation for all" by 2030. These goals also include a variety of sub-goals such as: "achieve access to adequate and equitable sanitation and hygiene for all and end open defecation, paying special attention to the needs of women and girls and those in vulnerable situations" and "the participation of local communities for improving water and sanitation management". The achievement of sanitation goals would ultimately prosper and aid other goals such as objective under Goal nr. 3 to ensure healthy lives and endorsing security for all individuals. Within the scope of the formulated objectives, sanitation service, therefore, needs to be improved and to consider an integrated approach to fulfil not only the technical aspects regarding pathogen removal, environmental compliance and resource recovery, but also have the capacity to adapt into diverse social-cultural environments where the perception of acceptable sanitation practices is different (Afolabi et al., 2017). In essence, water and sanitation are intrinsically a vital part of many indicators across the entire array of SDGs.

1.2 Overview of the approaches to safeguarding people's health

The range of possible adaptive responses for safeguarding people's health was first identified in the 1970s and catalogued as a series of methods, policies and practices that governed the sanitation service chain, which was composed of various elements, as reported by Rosenqvist et al. (2016). The first approach to sanitation services started, viewing and aiming the focus of sanitation services away from income generation and towards the fulfilment of basic human needs. This particular concept was also seen as a way of protecting public health and transferring the responsibility for sanitation to the national level (Young et al., 1988). Subsequently, the World Bank and other international financial institutions and governments directed more resources to the cities for construction of centralised systems with combined and/or separate sewerage. However, the provision of sanitation service provided by centralised sanitation systems is challenging due to high investment requirements, which are difficult to obtain even today (Rosenqvist et al., 2016). For instance, the centralised sanitation alternatives involve high capital costs, which were estimated to be around 30 billion dollars in 2001 and are predicted to increase to 75 billion by the year 2025 (Esray, 2001). These values do not include the necessary operation and maintenance cost needed for the sewerage network. Approximately 80 to 90% of the capital and operational costs are linked to the expense of sewerage conveyance systems with economies of scale associated with its implementation in densely populated areas (Libralato et al., 2012). As stated by van Lier et al. (1999); "Apart from its proven benefits, a large sewage network, is nothing more than a transportation system for human excreta and/or industrial water with valuable drinking water as the transport medium". While centralised sewer systems will remain the best available option in many cases, a more critical approach is necessary to consider other viable alternatives, where applicable.

Many of the drawbacks associated with centralised systems, mentioned previously, are partially addressed by improving the sanitation coverage through on-site sanitation facilities. These facilities, which are currently being used by approximately 3.1 billion people around the World, include on-site storage, such as a flush or pour-flush toilets connected to a septic tank, and dry or wet pit latrines; thus, separating the faeces characterised with high pathogenic content from direct and indirect contact with the user, which in turns enable protection of the environment and public health (Connor et al., 2017). One of the comprehensive and rather complete overviews of on-site sanitation technologies is published by Swiss Federal Institute of Aquatic Science and Technology (Eawag) (Tilley et al., 2014; Gensch et al., 2018). The schematic representation of the faecal/municipal sludge management chain is shown in Figure 1.

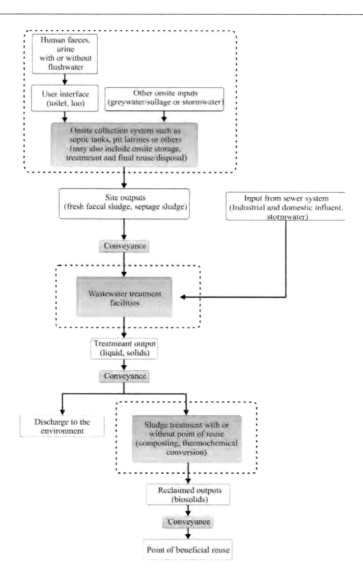

Figure 1: Schematic representation of the faecal and municipal sludge management chain (adapted from Mihelcic (2018)).

One-third of the population worldwide lacks access to basic sanitation, with the challenge of providing sanitation service in rapidly increasing urban areas that typically lack space or finances for centralised service level investment, as well as cheaper solutions (Mosello, 2016). The lack of provision of sanitation services and thus, the unsanitary living conditions observed in densely populated areas make public health hazards more prominent than typical urban settlements (Zakaria, 2019). Unsafe sanitation conditions, which favours the spread of diseases more rapidly, is a situation also commonly faced in a post-disaster situation, such as emergency relief sites, caused by the result of a human-made and/or natural disaster (Zakaria, 2019).

Furthermore, it is generally accepted that the technical advances alone provide only a partial solution to inadequate sanitation and that sustainable solutions require innovations at all parts of the sanitation service provision chain. A recently formalised approach by the World Bank, called City-wide Inclusive Sanitation (CWIS) recognises the multifaceted nature of sanitation and states that it "…looks to shift the urban sanitation paradigm, aiming to ensure everyone has access to safely managed sanitation by promoting a range of solutions—both on-site and sewered, centralised or decentralised—tailored to the realities of the World's burgeoning cities. CWIS means focusing on service provision and its enabling environment, rather than on building infrastructure".

1.2.1 Approaches to sludge management

Humans produce excreta (urine and faeces) daily, ranging from 0.6 to 2.1 L of urine cap^{-1} day^{-1} and from 130 to 530 g of fresh excreta cap^{-1} day^{-1} (Rose et al., 2015; Afolabi et al., 2017). In case of centralised systems, excreta end up in the sewers and may eventually undergo treatment at centralised wastewater treatment plants. In the treatment process, so-called sewage sludge is produced and it needs further treatment before it is disposed of or reused. In case of decentralised systems, excreta may be captured, contained and (partially) treated on-site, and in some cases, emptied as faecal or septic sludge, and transported to a centralised treatment plant (Figure 1). Often, treated sludge (e.g., fertiliser, charcoal) or end-products (e.g., biogas, nutrients) are reused or disposed of.

The faecal/septic sludge management, or as some call it non-sewered sanitation, has been neglected for decades, and only recently, attention has been paid to it, both financially and in terms of education and research. It is important to note that in this thesis, a distinction has been made in terminology between fresh faecal sludge, septic sludge and sewage sludge. The definitions adopted here are as stipulated in the recent literature (Strande et al., 2014; Guang-Hao Chen, 2020; Velkushanova et al., 2021).

Methods for the treatment of various sludges are numerous, and costly to the extent that presents a major investment and operational cost. Since its early developments, experts are trying to bring down the cost of treatment per unit of sludge (a ton of dry solids – DS, or a ton of 'wet' sludge), and until today the cost reduction remains the challenge.

Sludge contains water. Depending on the type, water content in the raw sludge can range from around 98%, for untreated sewage sludge or dilute septic sludge, to approximately 70-75%, for partially dewatered sewage sludge or some thicker septic sludge and fresh faecal sludge. This consequently means 2 and 25-30% DS, respectively. A specific distinction is applied among faecal sludge professionals where four types of sludge are recognised, depending on its DS%, namely: liquid (DS<5%), slurry (DS =5-15%), semi-solid (DS=15-25%) and solid (DS>25%) faecal sludge (Velkushanova et al., 2021).

In general, faecal sludge treatment objectives are stabilisation, nutrient management, pathogen inactivation and dewatering/drying (Niwagaba et al., 2014). The focus of this research is the treatment of sludge in terms of reduction of volume by drying. In principle, water in sludge is present in four distinctive forms, some more difficult to remove than others. This aspect will be addressed in the following section.

Methods for dewatering and drying or their combination can be divided into low-rate (slow, extensive) or high-rate (rapid, intensive). The purpose of sludge drying is to create end-product, solids that can be reused, most frequently as fuel. Technological options range from non-carbonised to carbonised, as presented in Figure 2. These technologies are well-capable of a high degree of drying, but the ultimate goal is still (how) to reduce the energy input per unit of sludge (kg DS or kg wet sludge).

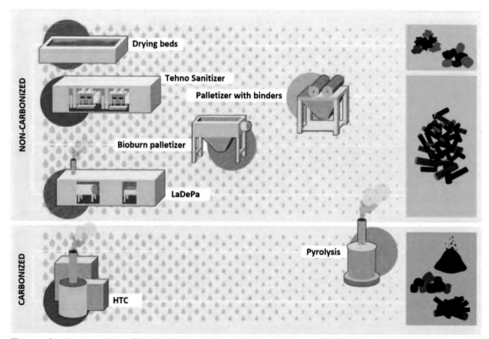

Figure 2: An overview of technology options for producing solid fuel, starting from dewatered faecal sludge at 80% moisture and ending at non-carbonised or carbonised solid-fuel end products. The position of the technology icons from left to the right indicates the required dryness of the input sludge for each technology, as indicated by the size of the droplets, ranging from 80% moisture on the left to 10% moisture on the right (figure and caption adopted from Velkushanova et al. (2021) which is modified from Andriessen et al. (2019).

The fundamental characteristics of sludge structure define the interaction between sludge and water and choice of technology for its removal. As said earlier (Figure 3), the water present in the sludge may exist in different physical forms, which are distinguished by their distribution, and by the intensity of their physical bonds ranging from capillary forces, adsorptive and adhesive forces to chemisorption and include (Neyens et al., 2003a): i) free water: water not attached to the solid particles or affected by capillary forces (void water), ii) interstitial moisture: water confined within crevices and capillaries of the sludge floc and organisms, iii) surface-water: liquid adsorbed or adhered to the surface of the solid particle and iv) intracellular and chemically bond water. Free water and water loosely attached to the pores and interstitial spaces of the sludge particles and flocs may be removed by using gravity or by application of external pressure (Flaga, 2005). Further removal of water characterised with a higher binding strength such as surface water, intracellular water and others may be achieved by applying heat-driven processes, such as drying technologies (Flaga, 2005).

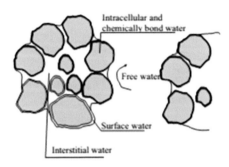

Figure 3: Distribution of water in the sludge floc (Neyens et al., 2003b)

In this thesis, dewatering is defined as a process used to remove free water, while drying is a process used to remove both free and bound water from the solid fraction of sludge.

As shown in Figure 2, traditional approaches to faecal sludge drying include sludge drying beds, which are available in a variety of options, such as sand drying beds, paved drying beds, planted drying beds, and many more (Wang et al., 2007; Nikiema et al., 2014). The sludge moisture reduction in sludge drying beds is achieved through percolation/filtration and evaporation process. The essential advantage of this technique, as illustrated in Figure 4, is the independence of the process from electrical energy, i.e., no energy is needed for the dewatering nor the drying process (Wang et al., 2007; Nikiema et al., 2014). However, the operation of the sludge drying beds is associated with relatively low drying rates (e.g., 0.06 kg m^{-2} h^{-1}), long residence time (weeks to months) and increased physical footprint requirements, which may be high as 0.08 m^2 capita^{-1} land requirement (Clark, 1970; Cofie et al., 2006; Nikiema et al., 2014). Consequently, sludge drying beds are not a viable option in areas characterised with space footprint constraints commonly observed in congested cities, slums and others (Ronteltap et al., 2014). The space footprint requirements of drying systems may be significantly reduced by the addition of auxiliary equipment such as ventilators, belt conveyors and/or by the introduction of external heat or energy sources such as hot air, heated oil, steam or, as in the case of LaDePa technology, infrared irradiation panels. In the LaDePa process and in other traditional drying approaches, the heat required for water vaporisation is initially delivered to the material's surface, which progressively spreads into the solid, mainly, through conduction mechanisms (Mujumdar et al., 2000; Bennamoun et al., 2013; Mujumdar, 2014; Septien et al., 2018). Depending on the heat source and its intensity, the heat transfer may occur rapidly, with a corresponding increase sludge drying rates, which may be high as 35 kg m^{-2} h^{-1} (Bennamoun et al., 2013). The drying techniques, mentioned above, may also be combined with extrusion and compaction processes supported by or without binding agents (Figure 2); thereby increasing the marketability of sludge-based products and thus generating revenues that could potentially offset the sludge drying costs (Nikiema et al., 2014; Andriessen et al., 2019). In other words, due to the increase in mechanical complexity and the high sensible and latent energy requirements of wet sludge, the specific energy required for the removal of water from solids, as shown in Figure 4, may be as high as 5 MJ L^{-1} (i.e., 1.4 kWh L^{-1}). Given the high specific energy consumption demand, pre-treatment technologies such as hydrothermal carbonation techniques can substantially improve the drying process's energy efficiency. The latter is directly related to the reduction in evaporative latent heat losses typically seen in conventional drying methods, accounting for energy costs of approximately 2.3 MJ kg^{-1} (i.e., 0.6 kWh kg^{-1}) (Bryan, 1907; Moon et al., 2015). However, as mentioned previously, to increase the sludge DS content up to 90%, post-treatment is required, including mechanical dewatering (70% DS) and drying techniques (90%DS) (Moon et al., 2015). Alternative options to pre-treatment processes include implementation of the cogeneration system, such as Rankine cycle (Sohail et al., 2018). During this process, the dry sludge (<90 DS%) or other fuel such as coal is combusted in a flue gas application unit to generate high-pressure steam that is converted to electricity by a reciprocating piston steam engine (Sohail et al., 2018). The process heat generated by the combustion process is used to dry the incoming faecal or septic sludge, which may be thereafter used as a fuel agent. One of the recently implemented technologies utilising such principle is the Janicki Omni-Processor (Sohail et al., 2018). The Janicki Omni-Processor is an energy-positive process while recovering water and nutrients from on-site sanitation facilities (Sohail et al., 2018). Other thermochemical processes, such as pyrolysis, are available to achieve the latter, which as opposed to combustion processes, occur within a limited supply of oxygen and can be used to produce several products, including oil, syngas and carbon (Barry et al., 2019; Nuagah et al., 2020). The resulting products can be applied to soil as a carbon supplement or used as an energy source that may be recuperated elsewhere. The carbonisation techniques, such as pyrolysis and hydrothermal carbonisation, are gaining momentum in sludge treatment. However, relatively limited research on carbonisation, specifically in terms of specific energy consumption requirements, excluded this technology in further considerations.

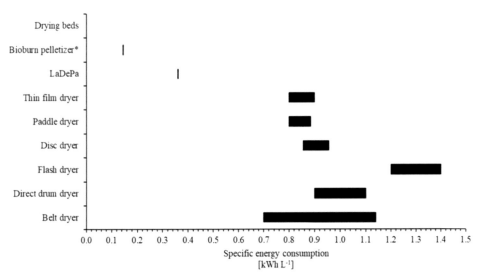

units in kWh kg⁻¹ of pellets

Figure 4: Specific energy consumption of different types of non-carbonised technologies (Léonard et al., 2011; Nikiema et al., 2014)

From the above overview and Figure 4, it can be concluded that the energy consumption for drying is still a critical factor for the technology selection. The drying techniques' energy demand may be reduced by implementing dewatering techniques, hydrothermal carbonisation techniques, and/or by reusing the energy released from the pyrolysis, (co-)combustion processes or other processes where the waste heat may be recuperated. Other contemporary developments have also been considered and researched by Mawioo et al. (2016a; 2016b; 2017). The authors suggested that microwave technology has potential in sludge treatment, ranging from faecal, septic and sewage sludge. However, the research of Mawioo and co-workers was not geared towards energy optimisation, because of the following reasons: (i) lack of thermal insulation of the drying system, (ii) inefficient use of microwave energy due to sub-optimized design of the reactor, (iii) less efficient mixing at higher sludge densities, (iv) low ambient temperature (5°C), (v) poor extraction of the condensate from the cavity, (vi) unnecessary heating of the cavity, and (vii) absence of energy recovery features (Kocbek et al., 2020; Velkushanova et al., 2021).

1.3 Microwave technology and research gaps

As introduced earlier in this chapter, the focus of this thesis is the use of microwave technology for sludge drying. Microwave technology is a relatively new technique successfully used by almost 1 billion people worldwide on a day-to-day basis at the household level, without requiring the user to have any knowledge of the basic principle of microwave functioning (Osepchuk et al., 2002). Microwave technology, however, offers an interesting range of application beyond the commonly known domestic purposes. Since 1945, the microwave has been successfully utilised for food processing (Venkatesh et al., 2004; Chandrasekaran et al., 2013), synthesis of organic and inorganic compounds (Abramovitch et al., 1991), pyrolysis of wood, (Borges et al., 2014; Wu et al., 2014), sewage sludge treatment (Menendez et al., 2002; Tyagi et al., 2013), devulcanization (Paulo et al., 2012) and processing of medical waste (Wu et al., 2008).

The microwaves are a form of non-ionising, electromagnetic waves with wavelengths from 1 to 1,000 mm and corresponding frequencies between 300 MHz and 300 GHz. Within this electromagnetic spectrum, frequencies are reserved for various applications in telecommunication, satellite communication, and radar, which can interfere with each other. Therefore, frequencies are allocated for industrial, scientific and medical (ISM) bands of 915 and 2,450 MHz, respectively (Thostenson et al., 1999; Bilecka et al., 2010). These frequencies cause interaction of dipolar molecules within particular materials and generate heat (Bhattacharya, et al., 2016). The alternating electric field energy positioned outside of absorbing material is irreversibly absorbed, resulting in quick 'volumetric"' heating with a conforming inverted temperature profile; i.e., upon exposure to microwave radiation, the microwaves penetrate the material, which generates heat, leading to quick sanitation and sludge mass/volume reduction. The volumetric heating effect has been reported to substantially enhance the system throughput capacity in sludge treatment systems, leading to rapid reductions in the sludge volume and mass, while simultaneously mitigating the chances of disease outbreaks related to excreta health threats through pathogen inactivation. Advantages of microwave drying applications also include fast process start-up and shutdown, lower adverse effects on global warming if renewable energy sources can produce the electrical energy needs for the microwaves, and low physical footprint (Maskan, 2000; 2001; Kouchakzadeh et al., 2010). Microwave thermal technology may therefore potentially pave the way for more compact and even mobile plants that can be deployed to rural areas or close to the end-user. That is why researchers considered microwave technology as a feasible sludge drying alternative that may provide treatment of both sewage sludge generated at centralised wastewater treatment facilities, and faecal sludge generated in remote decentralised areas without access to sewerage (such as informal settlements and/or emergency camps, (Afolabi et al., 2017; Mawioo et al., 2017). Accordingly, using this technology, treatment of sludge generated at wastewater treatment plants and on-site sanitation facilities, can be considered as a viable option (Mawioo et al., 2017).

However, despite the benefits of microwave technology, further research is required to optimise the value recovery and the efficiency of treatment by microwave heating processes, specifically on energy needs and process optimization. Current research shows that microwave technology possesses a comparatively high energy footprint (> 4 kWh L^{-1} or 16 MJ L^{-1}) when employed in sludge volume reduction in comparison with traditional drying methods, e.g., conductive and convective based dryers (Mawioo et al., 2017). Mitigation measures include introduction of a water vapour extraction unit that might ease the difficulties caused by internal cavity vapour condensation, introduction of turntable units that would allow the microwave energy to be more uniformly distributed within the sample, and isolation material additives for the reduction of the losses caused by the dissipated energy and the heat into the environment (Mawioo et al., 2017).

Alternative approaches that may raise the energy efficiency of the microwave performance in the treatment of sludge also include increase in sludge solids content prior the drying process (e.g. mechanical dewatering techniques), as well as the knowledge of the changes in a sludge's moisture content affected by the atmospheric conditions such as the relative humidity, temperature and pressure (Schaum et al., 2010; Bougayr et al., 2018).

All these measures can reduce the total energy demands for the unit and remove moisture more efficiently, thereby effectively reducing pathogenic content, sludge volume and mass.

Furthermore, the currently available research on microwave drying of sewage and faecal sludge is aimed at applying microwave technology for sludge at different power level, and exposure time, using different sample mass and type of sludge, aimed at pathogen destruction and sludge mass/volume reduction (Mawioo et al., 2016a; 2016b; 2017; Karlsson et al., 2019).

The evaluation of microwave technology in the treatment of sludge was mainly obtained by intermediate sampling, meaning that the sludge mass and temperature were monitored sequentially by discontinuing the operation of the system; hence the changes in the parameters monitored, for instance, in sludge mass cannot fully reflect the accuracy of the results due to potential losses in sludge moisture content occurring during the sampling period. Consequently, the quality of the data gathered is rather insufficient to allow thorough fundamental understanding, and factors that provide essential information regarding the way materials behave during drying are not accurately depicted and understood.

Furthermore, in order to enable microwave treatment *in-situ* of sludge, several other factors need to be considered, including; i) implementation of measures that may reduce the costs of drying (e.g., mechanical dewatering unit) and ii) incorporation of ancillary water treatment technologies, such as ultrafiltration and reverse osmosis that may enable the integration of treatment of the liquid streams generated from the mechanical and drying process.

As such microwave technology, when coupled with complementary technologies, can increase the process reliability treating sludge with high water content and pathogenic organisms and thus guarantee that the end products could be reused and pollution reduced.

1.4 Research hypotheses and objectives

In this study, the microwave technology is proposed as one of the viable technological alternatives to conventional drying, to be applied as a centralised, or *in-situ* treatment options for municipal sewage sludge, fresh faecal sludge, or septic sludge derived from on-site sanitation facilities. This technology may be operated as a standalone treatment unit, or coupled with mechanical dewatering techniques, membrane separation technology and, thus potentially increase the reliability of technology in the treatment of sludge while recovering useful resource that may be recovered through an agricultural or thermochemical application such as (co-) combustion. Based on the research gaps outlined in the previous sections, this particular research focuses on the development and evaluation of a novel microwave-based sludge treatment system both as standalone treatment and in conjunction with mechanical dewatering and membrane separation technology.

The overall objective of this study is to address the challenges in curbing the pollution and protect public health against factors caused by exposure to and contamination by untreated sludge/wastewater through the application of a novel mobile/semi-decentralised treatment concept, with microwave technology at the heart of it.

This research addresses the following hypotheses:

i. The overall specific energy consumption reported in microwave treatment of sludge can be significantly reduced (up to 70%) in comparison to results of present studies, and this can be achieved by improving the design and construction of the microwave system, including the drying chamber, the microwave generator and the water vapour extraction unit. Once this is in place, the specific energy consumption of the system will be in the range of traditional drying technologies (e.g., approximately 1.0 kWh L^{-1} or 3.6 MJ L^{-1});

ii. The thickness of the sample limits the propagation of (microwave-induced) energy within the sample, and this limitation can be counteracted by the microwave-selective 'moisture levelling' effect[1];

iii. Variation in sludge composition (specifically concerning the distribution of water as the result of its fat and oil content), has a direct influence on the (efficiency of) operation of microwave-based dryers, and thus, on the operational cost and the design of the system;

iv. Integration of existing technologies for wastewater and sludge treatment, such as microwave technology coupled with mechanical dewatering and membrane separation technology, enables efficient treatment of various types of sludge *in-situ*.

The specific research objectives are as follows:

i. To identify and implement appropriate solutions to address the microwave energy performance issues associated with high specific energy consumption (SEC) observed so far in microwave sanitisation and drying of sludge[2];

ii. To evaluate and analyse the drying and energy performance of a novel microwave pilot-scale treatment system in the treatment of dewatered waste activated sludge as a function of microwave power output[2];

---

1 Microwave 'moisture levelling' effect selectively improves microwave absorption rates in material regions characterized by a high dielectric loss.

2 This specific objective is addressing the hypothesis nr. i

iii. To compare the energy performance of the microwave treatment system in terms of specific energy consumption with the results of present studies carried out on microwave drying of sludge and conventional sludge drying treatment systems[2];

iv. To study the drying and energy performance of a microwave pilot-scale system in the treatment of municipal dewatered activated sludge at various thickness levels ranging from 45 to 150 mm at constant sludge sample mass[3];

v. To determine the microwave penetration limits and moisture levelling effect by measuring the temperature at various thickness level in the samples in unit of time[3];

vi. To develop a fundamental understanding of the impact of microwave operating conditions (such as microwave power output, sludge dimensions and mass) on the microwave system energy and throughput performance in the treatment of municipal sludge[2,3];

vii. To measure the physical-chemical properties (such as porosity, moisture, organic, fat and oil content, heavy metals, calorific value) and sorption isotherms for the selected sludge types, including samples from municipal wastewater treatment plant and on-site sanitation facilities[4];

viii. To evaluate the drying and energy performance of the microwave treatment system in treatment of different types of sludge[4];

ix. To determine if a relationship exists between the sorption isotherms, the fat and oil content of sludge, drying and energy performance of the system in treatment of various types of sludges[4];

x. To evaluate and assess the performance and applicability of technologies (such as mechanical dewatering, microwave irradiation, ultrafiltration and reverse osmosis) in treatment of septic tank sludge (sludge from septic tanks) under field-testing conditions in Jordan Valley (Jordan)[5].

1.5 Research approach

In order to achieve the research objectives listed above, the study was aimed, firstly, at assessing the microwave technology performances so far used in the treatment of sludge, both in the laboratory and pilot-scale applications. Special attention was placed towards the evaluation of technological problems that were associated with low (energy) efficiency of the system and identifying solutions to improve it.

With the knowledge obtained from scientific literature and experiences from industrial partners, a pilot microwave plant was designed and built. Following the manufacturing of the unit, a fundamental understanding of the sludge drying process under various microwave operating conditions and an insight into suitability and effectiveness of the proposed design for its

---

3 This specific objective is addressing the hypothesis nr. ii
4 This specific objective is addressing the hypothesis nr. iii
5 This specific objective is addressing the hypothesis nr. iv

implication in the demo pilot was investigated. The results obtained from the pilot-scale testing allowed us to make informed choices on the necessary modification required to enhance the efficiency of the unit in terms of specific energy consumption and maybe used for future reference to upscale the system and to stir the research towards other design alternatives.

Alongside the effect of operational parameters on the treatment of sludge using microwave technology, the effect of different types of sludge, and as such physical-chemical properties of sludge, on microwave performance was assessed. Emphasise was placed on the evaluation of water sorption properties and heat of sorption of different types of sludges tested. The aforementioned parameters provide essential information on the ability of the material to absorb or desorb water from the environment and the energy required to remove the water at different stages of drying. Thus, the analysis of water vapour sorption properties and heat of sorption of sludge may be used for the design of a drying process and/or identification of the necessary condition for sludge storage and thus final processing sludge moisture content. Finally, the synergistic effect of the technologies incorporated in the full-scale mobile system was investigated in Jordan.

The system made effective use of the electromagnetic energy derived from the operation of microwave technology, the energy gained from the mechanical action of the dewatering unit and physical barrier provided by ultrafiltration and reverse osmosis membrane. The system was evaluated by assessing the characteristics of obtained outputs and determining the mass balances and efficiencies of the monitored technologies. The evaluation of the system is also aimed at assessing the flexibility of the microwave system in terms of treatment capability, which was adapted for sanitisation of the sludge (pathogen reduction) and the production of the end product with high calorific value (sludge volume reduction).

1.6 Outline of the thesis

This thesis is divided into six chapters. The graphical representation of the thesis structure and connection of the chapters with the hypothesis of this doctoral thesis is shown in Figure 5. The first chapter provides a general introduction, knowledge gaps and rationale for the research topic. The second chapter deals with the assessment of design solutions implemented in a novel microwave-based pilot scale evaluated by quantifying and monitoring the energy consumption, sludge drying rates and the exposure time required to reduce the sludge volume and mass. The performance indicators used to evaluate the success of the microwave pilot plant system developed in this study were further used to illustrate the technical and economic potential of this technology in the treatment of sludge affected by operating parameters such as microwave power output (Chapter 2), sludge thickness and mass (Chapter 3) and physicochemical properties of sludge (Chapter 4). The fundamentals of sludge volume/mass loss and sanitisation under the influence of microwave operational parameters are discussed throughout the study. The fifth chapter evaluates and presents the findings of the field-testing assessment of the microwave-based containerised mobile system assisted with a wide range of complementary technologies such as mechanical dewatering unit and membrane separation technology. The final chapter concludes and summarises the findings of this research and provides further recommendations and perspectives on the subject.

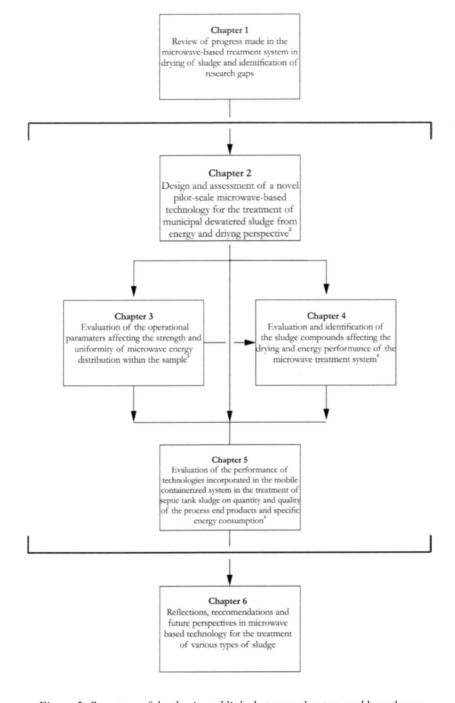

Figure 5: Structure of the thesis and links between chapters and hypotheses

References

Abramovitch RAJOp, international p. (1991). Applications of microwave energy in organic chemistry. A review. 23(6), 683-711.

Afolabi OO, Sohail M. (2017). Microwaving human faecal sludge as a viable sanitation technology option for treatment and value recovery–A critical review. Journal of Environmental Management, 187, 401-415.

Andriessen N, Ward BJ, Strande LJJoW, Sanitation, Development Hf. (2019). To char or not to char? Review of technologies to produce solid fuels for resource recovery from faecal sludge. 9(2), 210-224.

Barry D, Barbiero C, Briens C, Berruti FJB, bioenergy. (2019). Pyrolysis as an economical and ecological treatment option for municipal sewage sludge. 122, 472-480.

Bennamoun L, Arlabosse P, Léonard A. (2013). Review on fundamental aspect of application of drying process to wastewater sludge. Renewable and Sustainable Energy Reviews, 28, 29-43.

Bilecka I, Niederberger M. (2010). Microwave chemistry for inorganic nanomaterials synthesis. Nanoscale, 2(8), 1358-1374.

Borges FC, Du Z, Xie Q, Trierweiler JO, Cheng Y, Wan Y, Liu Y, Zhu R, Lin X, Chen P, Ruan R. (2014). Fast microwave assisted pyrolysis of biomass using microwave absorbent. Bioresource Technology, 156, 267-274.

Bougayr EH, Lakhal EK, Idlimam A, Lamharrar A, Kouhila M, Berroug F. (2018). Experimental study of hygroscopic equilibrium and thermodynamic properties of sewage sludge. Applied Thermal Engineering, 143, 521-531.

Bryan GH. (1907). Thermodynamics: An Introductory Treatise Dealing Mainley with First Principles and Their Direct Applications (Vol. 21): BG Teubner.

Chandrasekaran S, Ramanathan S, Basak T. (2013). Microwave food processing—A review. Food Research International, 52(1), 243-261.

Clark EE. Water treatment sludge drying and drainage on sand beds. University of Massachusetts, 1970.

Cofie OO, Agbottah S, Strauss M, Esseku H, Montangero A, Awuah E, Kone D. (2006). Solid–liquid separation of faecal sludge using drying beds in Ghana: Implications for nutrient recycling in urban agriculture. Water Research, 40(1), 75-82.

Connor R, Renata A, Ortigara C, Koncagül E, Uhlenbrook S, Lamizana-Diallo BM, Zadeh SM, Qadir M, Kjellén M, Sjödin JJTUNWWDR. (2017). The United Nations world water development report 2017. Wastewater: The untapped resource.

Corcoran E. (2010). Sick water?: the central role of wastewater management in sustainable development: a rapid response assessment: UNEP/Earthprint.

Esray SA. (2001). Towards a recycling society: ecological sanitation-closing the loop to food security. Water Science and Technology, 43(4), 177-187.

Flaga A. (2005). *Sludge drying*. Paper presented at the Proceedings of Polish-Swedish seminars, Integration and optimization of urban sanitation systems. Cracow March.

Gensch R, Jennings A, Renggli S, Reymond PJGWN, Swiss Federal Institute of Aquatic Science, Technology GWC, Sustainable Sanitation Alliance . Berlin G. (2018). Compendium of Sanitation Technologies in Emergencies.

Guang-Hao Chen MCMvL, G.A. Ekama, Damir Brdjanovic. (2020). Biological Wastewater Treatment (2nd Edition ed.).

Hutton G, Haller L, Bartram JJJow, health. (2007). Global cost-benefit analysis of water supply and sanitation interventions. 5(4), 481-502.

Karlsson M, Carlsson H, Idebro M, Eek C. (2019). Microwave heating as a method to improve sanitation of sewage sludge in wastewater plants. IEEE Access, 7, 142308-142316.

Kocbek E, Garcia HA, Hooijmans CM, Mijatović I, Lah B, Brdjanovic D. (2020). Microwave treatment of municipal sewage sludge: Evaluation of the drying performance and energy demand of a pilot-scale microwave drying system. Science of The Total Environment, 742, 140541.

Kouchakzadeh A, Shafeei S. (2010). Modeling of microwave-convective drying of pistachios. Energy Conversion and Management, 51(10), 2012-2015.

Léonard A, Crine M, Arlabosse P. (2011). Energy efficiency of sludge drying processes.

Libralato G, Ghirardini AV, Avezzù FJJoem. (2012). To centralise or to decentralise: An overview of the most recent trends in wastewater treatment management. 94(1), 61-68.

Maskan M. (2000). Microwave/air and microwave finish drying of banana. Journal of Food Engineering, 44(2), 71-78.

Maskan MJJofe. (2001). Drying, shrinkage and rehydration characteristics of kiwifruits during hot air and microwave drying. 48(2), 177-182.

Mawioo PM, Garcia HA, Hooijmans CM, Velkushanova K, Simonič M, Mijatović I, Brdjanovic D. (2017). A pilot-scale microwave technology for sludge sanitization and drying. Science of the Total Environment, 601, 1437-1448.

Mawioo PM, Hooijmans CM, Garcia HA, Brdjanovic D. (2016a). Microwave treatment of faecal sludge from intensively used toilets in the slums of Nairobi, Kenya. Journal of Environmental Management, 184, 575-584.

Mawioo PM, Rweyemamu A, Garcia HA, Hooijmans CM, Brdjanovic D. (2016b). Evaluation of a microwave based reactor for the treatment of blackwater sludge. Science of the Total Environment, 548, 72-81.

Menendez J, Inguanzo M, Pis J. (2002). Microwave-induced pyrolysis of sewage sludge. Water research, 36(13), 3261-3264.

Mihelcic JJJR. (2018). Sludge Management: Biosolids and Fecal Sludge.

Moon J, Mun T-Y, Yang W, Lee U, Hwang J, Jang E, Choi C. (2015). Effects of hydrothermal treatment of sewage sludge on pyrolysis and steam gasification. Energy Conversion and Management, 103, 401-407.

Mosello B. (2016). Sanitation Under Stress: How Can Urban Services Respond to Acute Migration? : Overseas Development Institute.

Mujumdar AS. (2014). Handbook of industrial drying (4th ed.): CRC press.

Mujumdar AS, Devahastin S. (2000). Fundamental principles of drying. Exergex, Brossard, Canada, 1(1), 1-22.

Neyens E, Baeyens J. (2003a). A review of thermal sludge pre-treatment processes to improve dewaterability. Journal of hazardous materials, 98(1-3), 51-67.

Neyens E, Baeyens J. (2003b). A review of thermal sludge pre-treatment processes to improve dewaterability. Journal of Hazardous Materials, 98(1–3), 51-67.

Nikiema J, Cofie OO. (2014). Technological options for safe resource recovery from fecal sludge.

Niwagaba CB, Mbéguéré M, Strande L. (2014). Faecal sludge quantification, characterization and treatment objectives. In Faecal sludge management: systems approach for implementation and operation (pp. 19-44): IWA Publishing London.

Nuagah MB, Boakye P, Oduro-Kwarteng S, Sokama-Neuyam YAJJoA, Pyrolysis A. (2020). Valorization of faecal and sewage sludge via pyrolysis for application as crop organic fertilizer. 151, 104903.

Osepchuk JMJIToMT, Techniques. (2002). Microwave power applications. 50(3), 975-985.

Paulo GD, Hirayama D, Saron C. (2012). Microwave devulcanization of waste rubber with inorganic salts and nitric acid. Paper presented at the Advanced Materials Research.

Ronteltap M, Dodane P-H, Bassan MJFSM-SAI, Operation. IWA Publishing L, UK. (2014). Overview of treatment technologies. 97-120.

Rose C, Parker A, Jefferson B, Cartmell E. (2015). The characterization of feces and urine: a review of the literature to inform advanced treatment technology. Critical reviews in environmental science and technology, 45(17), 1827-1879.

Rosenqvist T, Mitchell C, Willetts J. (2016). A short history of how we think and talk about sanitation services and why it matters. Journal of Water, Sanitation and Hygiene for Development, 6(2), 298-312.

Schaum C, Lux J. (2010). Sewage sludge dewatering and drying. ReSource–Abfall, Rohstoff, Energie, 1, 727-737.

Septien S, Singh A, Mirara SW, Teba L, Velkushanova K, Buckley CA. (2018). 'LaDePa' process for the drying and pasteurization of faecal sludge from VIP latrines using infrared radiation. South African journal of chemical engineering, 25, 147-158.

Sohail M, Cavill S, Afolabi OO. (2018). Transformative technologies for safely managed sanitation. Paper presented at the Proceedings of the Institution of Civil Engineers-Municipal Engineer.

Strande L, Brdjanovic D. (2014). Faecal sludge management: Systems approach for implementation and operation: IWA publishing.

Thostenson E, Chou T-W. (1999). Microwave processing: fundamentals and applications. Composites Part A: Applied Science and Manufacturing, 30(9), 1055-1071.

Tilley E, Ulrich L, Lüthi C, Reymond P, Zurbrügg C. Compendium of sanitation systems and technologies. Eawag; 2014. Swiss Federal Institute of Aquatic Science and Technology (Eawag), Dübendorf, Switzerland, 2014.

Tyagi VK, Lo S-L. (2013). Microwave irradiation: A sustainable way for sludge treatment and resource recovery. Renewable and Sustainable Energy Reviews, 18, 288-305.

UN. World Urbanization Prospects. 2021, https://esa.un.org/unpd/wup/Download/. 2018.

UN PD, United Nations Department of Economic and Social Affairs. World population prospects 2019: highlights.2021,https://population.un.org/wpp/Publications/Files/WPP2019_Highlights.pdf., 2019.

van Lier JB, Lettinga GJWS, Technology. (1999). Appropriate technologies for effective management of industrial and domestic waste waters: the decentralised approach. 40(7), 171-183.

Velkushanova K, Strande L, Ronteltap M, Koottatep T, Brdanovic D, Buckley C. Methods for Faecal Sludge Analysis. 2nd Edition. IWA Publishing, London, UK, 2021.

Venkatesh M, Raghavan G. (2004). An overview of microwave processing and dielectric properties of agri-food materials. Biosystems Engineering, 88(1), 1-18.

Wang LK, Li Y, Shammas NK, Sakellaropoulos GP. (2007). Drying beds. In Biosolids Treatment Processes (pp. 403-430): Springer.

WHO WHO. Diarrhoeal disease, https://www.who.int/news-room/fact-sheets/detail/diarrhoeal-disease, 2017.

Wu C, Budarin VL, Gronnow MJ, De Bruyn M, Onwudili JA, Clark JH, Williams PT. (2014). Conventional and microwave-assisted pyrolysis of biomass under different heating rates. Journal of Analytical and Applied Pyrolysis, 107, 276-283.

Wu T-NJPPoH, Toxic,, Management RW. (2008). Environmental perspectives of microwave applications as remedial alternatives. 12(2), 102-115.

Young B, Briscoe J. (1988). A case-control study of the effect of environmental sanitation on diarrhoea morbidity in Malawi. Journal of Epidemiology & Community Health, 42(1), 83-88.

Zakaria F. (2019). Rethinking Faecal Sludge Management in Emergency Settings: Decision Support Tools and Smart Technology Applications for Emergency Sanitation: CRC Press.

2

Microwave treatment of municipal sludge

Drying performance and energy demand of a novel pilot-scale microwave drying system

This chapter is based on: Kocbek E, Garcia HA, Hooijmans CM, Mijatović I, Lah B, Brdjanovic D. Microwave treatment of municipal sewage sludge: Evaluation of the drying performance and energy demand of a pilot-scale microwave drying system. Science of The Total Environment 2020; 742: 140541.

Abstract

Sewage sludge management and treatment can represent up to approximately 30% of the overall operational costs of a wastewater treatment plant. Microwave (MW) drying has been recognized as a feasible technology for sludge treatment. However, MW drying systems exhibit high energy expenditures due to: (i) unnecessary heating of the cavity and other components of the system, (ii) ineffective extraction of the condensate from the irradiation cavity, and (iii) an inefficient use of the microwave energy, among others issues. This study investigated the performance of a novel pilot-scale MW system for sludge drying, specifically designed addressing the shortcomings previously described. The performance of the system was assessed drying municipal centrifuged wasted activated sludge at MW output powers from 1 to 6 kW and evaluating the system's drying rates and exposure times, specific energy outputs, MW generation efficiencies, overall energy efficiencies, and specific energy consumption. The results indicated that MW drying significantly extends the duration of the constant rate drying period associated with the evaporation of the unbound sludge water, a phase associated with low energy input requirement for evaporating water. Moreover, the higher the MW output power, the higher the sludge power absorption density, and the MW generation efficiency. MW generation efficiencies of up to 70% were reported. The higher the power absorption density, the lower the chances for energy losses in the form of reflected power and/or energy dissipated into the MW system. Specific energy consumptions as low as 2.6 MJ L^{-1} (0.74 kWh L^{-1}) could be achieved, well in the range of conventional thermal dryers. The results obtained in this research provide sufficient evidence to conclude that the modifications introduced to the novel pilot-scale MW system mitigated the shortcomings of existing MW systems, and that the technology has great potential to effectively and efficiently dry municipal sewage sludge.

2.1 Introduction

Municipal sewage sludge is a by-product of the treatment of municipal and industrial wastewater. Depending on the source of the wastewater, the sewage sludge may contain pathogenic organisms, antibiotic-resistant microorganisms, and inorganic and organic pollutants such as polycyclic aromatic hydrocarbons, dioxins, furans, heavy metals, and pharmaceutical compounds, among others (Raheem et al., 2018). Consequently, the disposal of sewage sludge may pose a risk to the environment and human health. Due to these concerns, the direct utilisation of sludge for agricultural activities has been limited or banned in many countries especially in Western Europe. Among the limited disposal options available, (co-)incineration is emerging as the most viable alternative for the final disposal of sludge (Kacprzak et al., 2017). However, it comes at a higher cost than other conventional treatment options such as the reuse of stabilised sludge in agriculture. According to Jakobsson (2014), the cost of (co-)incineration varies from between 250 to 330 EUR per tonne of dry solids (DS), which is much higher than the cost of the sludge treatment (stabilisation) commonly used to enable sludge to be reused in agriculture (160 to 210 EUR per tonne of DS). A major factor contributing to the fluctuations in sludge treatment costs is the cost of transporting the sludge from its source to the treatment and final disposal locations. To achieve economies of scale, sludge (co-)incineration facilities are often centralised installations serving more than one wastewater treatment plant (WWTP) in densely populated areas. Such sludge incineration facilities usually charge between 80 and 120 EUR per tonne of wet sludge (Kacprzak et al., 2017). Sludge management and treatment can represent up to approximately 30% of the total operational costs of a WWTP (Jakobsson, 2014).

Sludge originating from municipal WWTPs consists primarily of water (98-99%). Therefore, most of the sludge treatment interventions (and related costs) are associated with large volumes of water requiring frequent collection, transport, and treatment. Typically, most of the water content is reduced using mechanical and/or thermal dewatering processes. Mechanical dewatering technologies utilise gravity or the application of external pressure to achieve their ends Vesilind (1994); (Schaum et al., 2010). Mechanical dewatering technologies such as centrifuges, decanters, belt filter presses, hydraulic filter presses, and screw presses can reduce the sludge moisture content by up to approximately 70% (i.e., 30% DS) (Schaum et al., 2010). Further removal of water can be achieved by applying conventional thermal processes involving the use of heated air, steam, or flue gas in belt conveyors, fluidised beds, spray dryers, or drums (Schaum et al., 2010). However, there are drawbacks to all of these conventional drying processes including low drying efficiencies, long exposure/treatment times, and substantial energy requirements (Ohm et al., 2009; Mujumdar, 2014).

Microwave (MW) radiation is proposed as a promising alternative for sludge treatment (Afolabi et al., 2017; Mawioo et al., 2017). MWs are a form of nonionizing electromagnetic waves commonly operated at frequencies of 915 and 2450 MHz (Haque, 1999; Bilecka et al., 2010). These frequencies are widely used in MW drying applications. The main mechanism of the conversion of nonionizing electromagnetic energy into heat during MW irradiation of sludge is the dipolar polarisation mechanism (Stuerga, 2006). The torque applied by the electric field induces the rotational motion of all molecules exhibiting a permanent dipole moment such as the water molecules within the sludge (Stuerga, 2006). The water molecules try to resist the changes caused by the oscillating field, producing elastic, inertial, frictional, and molecular interaction forces and resulting in a temperature increase throughout the material (Mishra et al., 2016). That is, the transformation of energy into heat is attributed to the ability of the electromagnetic energy to couple and induce the polarisation of charges inside the irradiated material. This ability is governed by the dielectric loss tangent (tan δ) of the material, that is, the ratio between the dielectric loss factor (ε'') and the dielectric constant (ε').

The potential application of MW radiation for sludge heating, including drying and sterilisation, has been successfully demonstrated in a laboratory setting. For example, a study carried out by Dominguez et al. (2004) on MW heating of sewage sludge showed a 10-times reduction of the exposure time compared with conventional heating technologies. Specifically, a conventional hot air furnace operating at 2 kW required 55 minutes to reduce the moisture content to less than 1% while a MW unit operating at 1 kW required just 5 to 8 minutes. Similarly, MW technology has shown promising results for pathogen inactivation in sludge. Hong et al. (2006) investigated the effects of both MW and conventional heating systems (water bath) on faecal coliform removal in municipal primary sludge. The MW radiation system reduced the number of faecal coliforms in the sludge below the detection limit at a temperature of 65 °C within a minute. The same results were achieved by the conventional heating system but at a higher treatment temperature of 100 °C and after a longer exposure time of 4.8 minutes. In addition to the inactivation of faecal coliforms, MW sanitisation of sludge has also been proven to be effective for inactivating other microorganisms such as *E. coli*, *Ascaris lumbricoides* egg, *Staphylococcus aureus*, and *enterococcus faecalis* below detection limits (Mawioo et al., 2016a; 2016b; 2017). The reduced processing time observed for MW treatment systems compared with conventional drying techniques may be due to the selective and penetrating nature of MWs. MW radiation generates heat from within the material being heated and causes heat generation throughout the volume of the material, resulting in a heating profile that emanates from the inside to the outside of the material. The generated heat caused by molecular friction leads to internal evaporation and the corresponding generation of internal pressure (Ni et al., 1999; Fu et al., 2017). Such pressure induces moisture to move to the surface of the material where it is then removed. Using MW radiation, a much higher flowrate (up to 10 times) of water from the inside of the material to the surface has been reported compared with convection drying processes (Kumar et al., 2016). Due to the selective heating properties of MW radiation, only dielectric materials such as water can be heated. Other materials such as teflon, polypropylene, and bulk metals are largely unaffected by MW radiation (Bhattacharya et al., 2016). Therefore, MW selective heating is one of the most advantageous attributes of MW technology as it reduces the energy inefficiencies associated with the indirect heating of the atmosphere, the surface of the MW irradiation cavity, or other components of the system commonly observed with conventional heating systems (Kappe et al., 2012). Thus, MW technology can offer an effective, fast, and flexible treatment option for the sanitisation and drying of sludge.

Despite the potential benefits of using MW radiation for sludge treatment, most of the evaluations of this technology have been carried out in a laboratory setting. To date, only limited attempts to scale-up testing have been made. Mawioo et al. (2017) evaluated the performance of a pilot-scale

MW reactor treating centrifuged waste activated sludge (C-WAS) and reported promising results regarding the sterilisation and drying of the sludge. However, the authors reported very high specific energy consumption (SEC) values (the energy consumed per litre of evaporated water) of 14.5 MJ L^{-1}. That is, Mawioo et al. (2017) reported SEC values that were three to six times higher than those of conventional thermal convective systems such as belt driers, direct drum dryers, and flash dryers (Léonard et al., 2011). Mawioo et al. (2017) identified several issues with the MW pilot system that could lead to such low SEC values: (i) poor design of the extraction of the condensate (water vapour) from the irradiation cavity, (ii) the occurrence of water condensation inside the MW irradiation cavity, (iii) cold ambient (winter) temperature, (iv) cold start-up of the system, (v) lack of thermal insulation, (vi) inefficient use of the microwave energy, (vii) unnecessary heating of the irradiation cavity, (viii) uneven distribution of the MW energy on the irradiated sludge, and (ix) absence of energy recovery features. Other researchers have also reported similar and additional factors that may contribute to such energy inefficiencies including the non-uniform distribution of both the electromagnetic field and temperature within the irradiated material and overall failures in the design of the MW system (Vadivambal et al., 2010). Previous research into sludge with MW systems at the pilot-scale have included the use of low power magnetrons that have achieved MW generation efficiencies of approximately 50% (Mawioo et al., 2016a; 2016b; 2017). Treating sludge using a MW generator with a higher quoted conversion efficiency may reduce the sludge treatment time and energy consumption. Previous research on MW sludge treatment has also failed to evaluate the impact of the different MW power outputs on the optimisation of energy efficiencies to identify an optimum MW power output to sludge load ratio (Bermúdez et al., 2015). In addition, vapour extraction has been incorrectly implemented, and the uniformity of the MW electric field on the irradiated sample has not been emphasised. Some of these shortcomings have been overcome by combining MW radiation with hot air treatment (Fennell et al., 2014; Fennell et al., 2015). The water evaporated by the action of the MW radiation is removed by the introduced stream of hot air, avoiding condensation and the rewetting of the irradiated sample. Moreover, the addition of the hot air treatment can contribute to the uniform heating of the material, improving the energy efficiency of the process (Kumar et al., 2014b; Cravotto et al., 2017). Additionally, intermittent MW convective drying with hot air can be applied (introducing the MW energy as sequential pulses), which provides more uniform heating than MW radiation alone (Joardder et al., 2013; Kumar et al., 2014a; Kumar et al., 2014b; Kumar et al., 2016; Joardder et al., 2017; Welsh et al., 2017; Kumar et al., 2019). Several other options are available that can overcome non-uniform electric field distribution and the corresponding heating pattern within the samples such as the use of travelling wave applicators, and/or the introduction of moving parts in the irradiation cavity such as agitators or turntables (Boldor et al., 2005; Fennell et al., 2014; Fennell et al., 2015).

MW technology has thus shown potential advantages in terms of pathogen inactivation and drying compared with conventional drying technologies. However, several limitations associated with the energy efficiency of large-scale MW systems for sludge treatment remain and threaten the practicality of such systems. This research evaluated the performance of a pilot-scale MW reactor for the treatment of C-WAS. The pilot-scale MW system evaluated in this research was designed to address all the shortcomings described in the previous paragraph, notably: (i) improvements on the overall design of the MW irradiation cavity to achieve the uniform distribution of MW radiation while avoiding heating the irradiation cavity, (ii) the inclusion of a more efficient extraction system to allow the condensate to leave the irradiation cavity, and (iii) the provision of a MW generator with a high quoted MW generation efficiency to more efficiently use the MW energy. The treatment performance of the pilot-scale MW system was assessed at different MW output powers, which determined the overall treatment/exposure times, the sludge drying rates, the specific energy consumption, the drying energy efficiencies, and the overall MW system energy efficiencies.

2.2 Materials and methods

2.2.1 Sludge samples

C-WAS samples were collected from the WWTP located in Ptuj, Slovenia (Figure 6). Polymer was added at the WWTP to improve the dehydration of the sludge.

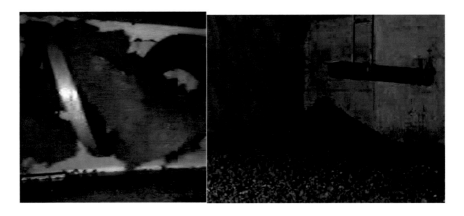

Figure 6: C-WAS sample collected at the WWTP located in Ptuj

2.2.2 Analytical procedures

2.2.2.1 Total solid (TS) and volatile solid (VS) determination

The TS and VS concentrations of the sludge were determined twice weekly according to the gravimetric methods SM-2540D and SM-2540E, respectively, as described in American Public Health et al. (2005). The sludge DS concentration was calculated from the TS concentration.

2.2.2.2 Calorific value and carbon, hydrogen, nitrogen, and sulphur determination

The calorific value of the raw sludge was measured in a bomb calorimeter (IKA- Calorimeter C 400 adiabatisch IKA®-Werke GmbH & Co. KG, Staufen German) according to the SIST-TS CEN/TS 16023:2014 standards. The elemental sulphur (S) and hydrogen (H) content in the sludge were measured using the Dumas method. The elemental total carbon (C) was determined using the dry combustion method according to SIST EN 13137:2002. The elemental total nitrogen (N) was determined according to SIST EN 16168:2013. Both the calorific value and the elemental analyses were determined at the Institute of Chemistry, Ecology, Measurement and Analytics (IKEMA, Lovrenc na Dravskem polju, Slovenia).

2.2.3 Experimental pilot-scale MW system

The pilot-scale MW system was designed and manufactured by Tehnobiro d.o.o (Maribor, Slovenia) for the treatment of diverse types of sludge with different water and solids content such as fresh faecal sludge, septic tank sludge, and waste sewage sludge. A detailed schematic drawing of the system and the general view of the pilot unit is presented in Figure 7. The MW system consisted of a stainless-steel cylindrical MW irradiation cavity provided with a polypropylene (PP) oval vessel with a maximum sludge load capacity of 6 kg and a holding vessel stand on a rotating PP turntable provided with an electromotor to rotate the sludge sample at a speed of 1 rpm. This rotational effect could potentially increase the uniform irradiation of the sludge. Ancillary equipment included a ventilation unit for the extraction of the condensate, a MW power supply, a MW magnetron with a maximum output power of 6 kW at a frequency of 2,450 MHz, and an air filtration system for odour control. The MW magnetron delivered the desired power along a standard rectangular waveguide WR340 (86.36 x 53.18 mm) connected to a circulator. An isolator connected to the MW head allowed the MW power to pass to the MW cavity (forward power) but not to flow in the reverse direction (reflected power). The reflected power was absorbed by a dummy load connected to the waveguide circulator. A teflon window was placed between the isolator outlet and the inlet of the MW chamber to prevent humidity, dust, and other elements that could damage the MW magnetron. A magnetron-cooling water-based system was incorporated to reduce the temperature in the MW magnetron and power supply. Demineralised water was provided at a flowrate of 600 L h^{-1}. Three different types of fillers were used to selectively increase the adsorption capacity of the air filtration system. These fillers were activated carbon soaked in phosphoric acid (H_3PO_4), activated carbon soaked in sodium hydroxide (NaOH), and aluminium oxide (Al_2O_3) with potassium permanganate (KM_nO_4). The changes in the sludge weight that occurred during the sludge's exposure to the MW drying process were continuously measured by a single point load cell (Mettler Toledo) with an error lower than 0.016 grams and a resolution of 5 grams. The electrical energy supplied to the MW unit was continuously measured using a power network analyser.

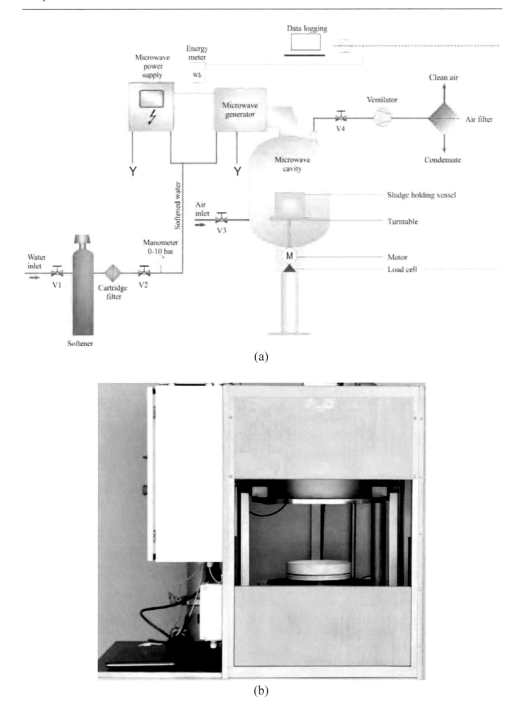

(a)

(b)

Figure 7: Schematic representation of the experimental pilot-scale MW system (a) and general view of the pilot system (b)

2.2.4 Experimental procedure

The C-WAS samples were collected on the same day prior to each evaluation. Fresh samples were weighed and placed in the PP holding vessels. The vessels were then placed on the rotating turntable inside the MW cavity. The experimental conditions for the evaluated samples are described in Table 1. The MW output powers were increased from 1 to 6 kW, while the initial sludge mass load and thickness remained constant at 3 kg and 60 mm, respectively. The experiments were finalised when the sludge moisture reached a value of approximately 0.18 kg of water kg of dry solid^{-1} (i.e., 85% DS). The evaluations were performed in either duplicate or triplicate and the average results were reported.

The evaluation of the MW pilot-scale unit was carried out in the research hall of the municipal WWTP located in Ptuj, Slovenia.

Table 1: Experimental design

Experiment No	Experimental parameters and variables		
	MW output power	Sludge mass	Sludge layer thickness
	[kW]	[kg]	[mm]
1	1.00	3.0	60
2	1.50	3.0	60
3	3.00	3.0	60
4	3.25	3.0	60
5	4.50	3.0	60
6	6.00	3.0	60

2.2.5 Data analysis

2.2.5.1 Specific energy consumption (SEC)

The SEC was calculated as shown in Equation 1:

$$SEC = \frac{P_{in;elect} \cdot t}{m_{eva}} \tag{1}$$

where the SEC is the specific energy consumption [kJ L^{-1} of water], $P_{in;elect}$ is the input power consumed by the system during the drying process [kW], t is the exposure time [s], and m_{eva} is the amount of evaporated water at a specific exposure time [L]. The $P_{in;elect}$ was measured using a power network analyser, as described in Section 2.2.3.

2.2.5.2 Specific energy output (SEO)

The SEO was calculated as shown in Equation 2 (Robinson et al., 2007):

$$SEO = \frac{P_{out,micr} \cdot t}{m_{sample}} \tag{2}$$

where the SEO is the specific energy output [kJ kg^{-1}], m$_{sample}$ is the initial mass of sludge (kg), t is the exposure time [s], and P$_{out;micr}$ is the output power (nominal power) supplied to the MW chamber [kW]. The MW output power was set according to the manufacturer's settings and ratings though the MW power supply control panel.

2.2.5.3 Specific energy input (SEI)

The SEI was calculated as shown in Equation 3:

$$SEI = \frac{P_{in;elect} \cdot t}{m_{sample}}$$
(3)

where the SEI is the specific energy input [kJ kg^{-1}].

2.2.5.4 Energy efficiency (μ_{en})

The μ_{en} is the ratio between the theoretical energy demand for evaporating the water and the energy consumed by the MW unit during the drying process, the calculation for which is shown in Equation 4 (Jafari et al., 2018):

$$\mu_{en} = \frac{(m_{sample} \cdot c_p \cdot \Delta T) + (m_{eva} \cdot h_{fg})}{P_{in;elect} \cdot t} \cdot 100$$
(4)

where μ_{en} is the energy efficiency [%], c_p is the specific heat capacity of the water [kJ kg^{-1} °C^{-1}], ΔT is the temperature difference of the sample between the exposure time t and the start of the treatment, and h$_{fg}$ is the latent heat of the water [kJ kg^{-1}] (2,257 kJ kg^{-1} at 100 °C as reported by (Haque, 1999)). The temperature of the samples was not measured; as such, the influence of sensible heat on total energy efficiency was considered assuming the sludge sample reached a temperature of 100 °C before water evaporation took place.

2.2.5.5 MW generation efficiency (μ_{gen})

The μ_{gen} is defined as the ratio between the MW energy output and the MW energy input, the calculation for which is shown in Equation 5 (Lakshmi et al., 2007):

$$\mu_{gen} = \frac{P_{out,micr} \cdot t}{P_{in;elect} \cdot t} \cdot 100$$
(5)

2.2.5.6 Sludge moisture content (X)

The X was calculated as shown in Equation 6 (Chen et al., 2014):

$$X = \frac{m_t - m_d}{m_d}$$
(6)

where X is the moisture content of the sludge [kg of water per kg of dry solid^{-1}] and m$_d$ is the total mass of dry solids in the sample [kg]. The total mass of dry solids in the samples was determined as the DS of the raw C-WAS sludge sample, as described in Section 2.2.2.1. The m_t was continuously determined by the point load cell located in the MW irradiation cavity, as described in Section 2.2.3. Therefore, X was continuously measured.

2.2.5.7 Drying rate (D_R)

The D_R was calculated as shown in Equation 7 (Chen et al., 2014):

$$D_R = \frac{dX}{dt} \tag{7}$$

where D_R is the drying rate [kg of water per kg of dry solid^{-1} min^{-1}] and X is the moisture content of the sludge at a specific exposure time. The drying rates were determined by polynomial regression analysis using Microsoft Excel and considered at time intervals of 30 s.

2.2.5.8 Power absorption density (P_d)

According to Maxwell equation, the power absorption density (i.e., the amount of power absorbed by a material per unit of volume) (P_d) is proportional to the input power, which relates to the electric field intensity (E), as shown in Equation 8 (Stuerga, 2006; Gupta et al., 2007):

$$P_d = 2\pi f \varepsilon_0 \varepsilon' |E|^2 \tag{8}$$

where P_d is the amount of absorbed power per unit volume [Wm^{-3}], f is the microwave frequency [s^{-1}], ε_0 is the permittivity of free space [8.85 × 10^{-12} Fm^{-1}] and E is the electric field intensity [Vm^{-1}]. The electric field intensity can be calculated as described in Equation 9 (Soltysiak et al., 2008; Pitchai et al., 2012).

$$E = \sqrt{\frac{2P_{out,micr}}{1 - |S_{11}|^2}} \tag{9}$$

where S_{11} is the reflection coefficient associated with the fraction of the power reflected by the sample.

2.3 Results and discussion

The effects of the different MW output powers on the drying performance of the C-WAS sludge were assessed and are presented in this section. The drying performance was evaluated by observing the effects of the MW output powers on exposure times, drying rates, specific energy consumption, drying energy efficiencies, and the overall energy efficiencies of the MW system.

2.3.1 Sludge physical-chemical properties

The raw C-WAS evaluated in this research exhibited the physical-chemical characteristics as presented in Table 2 below.

The sludge was generated by a sequencing batch reactor treatment process at a WWTP in Slovenia that receives domestic and industrial wastewater. Such sludge has a 2% DS concentration. The sludge is dewatered by centrifuges up to a DS concentration of approximately 17% (83% moisture content), as described in Table 2. The waste from several food processing facilities (mostly poultry slaughterhouses) discharge waste into the Ptuj WWTP, which explains the high organic content of its sludge (i.e., 88% volatile solids). The final DS concentration of 17% achieved by the centrifuges is in accordance with conventional DS concentration values between 13 to 21% (87 to 79% moisture content) reported in the literature for municipal sludge dewatering (Léonard et al., 2004; Mawioo et al., 2017). The gross calorific value of the sludge, as well as the elemental C, H, N, and S content, define the inherent energy content of the sludge, as well as the air pollution potential if the sludge is combusted. The dried sludge in this research exhibited a gross calorific value of 18.4 MJ kg^{-1} of DS. Such calorific value is in accordance with the calorific values of approximately 19 MJ kg^{-1} previously reported for dried sludge (Chen et al., 2014; Mawioo et al., 2017), as well as with calorific values for wood species such as birch, sapwood, and maple with calorific values ranging from 17.9 to 18.5 MJ kg^{-1} (Günther et al., 2012). The relatively high calorific value reported in this research was in accordance with the high elemental C content observed in the dried sludge of 46.5% (Table 2), also reported by Akdağ et al. (2018). The elemental composition of H, N, and S represented 5, 9.9, and 1.5% of the dried sludge, respectively (Table 2). Akdağ et al. (2018) reported similar findings related to the elemental H, N, and S concentrations of 3 to 5.3%, 1.88 to 7.7%, and 0.55 to 1.79%, respectively. The elemental N and S content on the dried sludge influences the formation of sludge combustion gasses such as NOx and SOx, which are harmful air pollutants. Thus, the sludge exhibited a high energetic value that can potentially be recovered as a source of energy. Nonetheless, the generation of combustion gasses with the potential to contribute to air pollution should be closely monitored.

Table 2: Physical-chemical characteristics of the raw C-WAS

Parameter	Unit	Value
Total solids (dry solids)	[%]	17 ± 1
Moisture content	[%]	83 ± 1
Volatile solids[a]	[%]	88 ± 2
C[a]	[%]	46.5
H[a]	[%]	5
N[a]	[%]	9.9
S[a]	[%]	1.5
Gross calorific value[a]	[MJ kg^{-1}]	18.4

[a]dry basis

2.3.2 Drying rates and exposure times

Figure 8 illustrates the sludge exposure time required to reduce the sludge moisture content from 4.88 to 0.18 kg of water kg of dry solids[-1] (i.e., from 17% to 85% DS content) at the evaluated MW output powers. The required exposure time decreased as the MW output power increased.

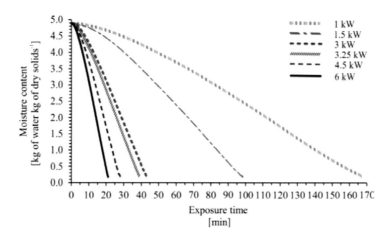

Figure 8: Effect of the MW output power on sludge drying as a function of the exposure time

The sludge drying rates were also calculated at each of the evaluated output powers and are presented in Figure 9 as a function of the exposure time. The sludge drying rates consistently increased with the MW output power across the evaluated range. As the MW output power increased from 1 to 6 kW, the maximum average sludge drying rates increased from 0.03 to 0.28 kg of water kg of dry solids[-1] min[-1]. Comparable sludge drying rates were reported in the literature on laboratory/bench-scale MW systems; however, the literature showed much lower sludge drying rates for pilot-scale MW systems. Bennamoun et al. (2016) and Chen et al. (2014) reported sludge drying rates of between 0.17 and 0.40 kg of water kg of dry solids[-1] min[-1] for a laboratory-scale MW system, whereas Mawioo et al. (2017) reported sludge drying rates of between 0.01 and 0.04 kg of water kg of dry solids[-1] min[-1] for a pilot-scale MW system.

Further, the sludge drying rates can be presented as a function of the sludge moisture content, as shown in Figure 10. Such plots (Krischer's plot) are commonly used to represent the drying process since they indicate the direct relation between the drying rates and the amount of water still present in the material. The abscissa in Figure 10 follows the opposite direction to the abscissa in Figure 9. That is, the higher the moisture content of the sludge, the lower the irradiation exposure time.

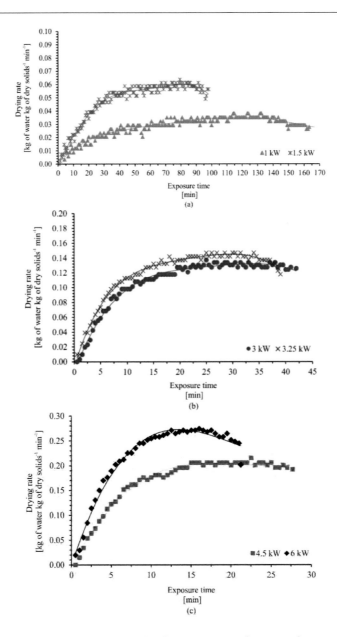

Figure 9: Effect of MW output power on the drying rate as a function of exposure time at MW output power of (a) 1 and 1.5 kW (b) 3 and 3.25 kW and (c) 4.5 and 6 kW

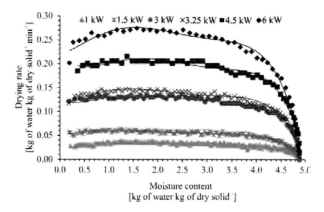

Figure 10: Effect of MW output power on the drying rate as a function of sludge moisture content

The changes in the sludge drying rates as a function of both the exposure time (Figure 8) and sludge moisture content (Figure 10) denote three main phases related to the drying process: (i) a drying rate adaptation period, (ii) a constant rate drying period, and (iii) a falling rate drying period. At the beginning of the adaptation period, the electromagnetic energy absorbed by the sludge resulted in a temperature increase with a subsequent increase of the sludge drying rate. The drying rate adaptation period is presented as a steep curve on the left-side and right-side of Figure 9 and Figure 10, respectively. The rapid increase in the sludge drying rate lasted for approximately $34 \pm 1\%$ of the exposure time at all the evaluated MW powers (Figure 9) up to a sludge moisture content of approximately 3.8 kg of water kg of dry solids^{-1} or 21% DS (Figure 10). After the adaptation period, the sludge drying rate remained largely constant until almost the end of the drying process. The constant rate drying period is associated with the removal of water molecules from the surface of the sludge. This water fraction is commonly described as unbound water or free water (Kopp et al., 2000). As a result, the evaporation of the water molecules located at the surface of the sludge (free water) determines the drying mechanism (and drying rates) during the constant rate drying period. The constant rate drying period lasts for as long as the rate at which the water molecules continue to be transported from the inside of the sludge to the surface of the sludge equals the rate at which the water molecules are evaporated from the surface of the sludge (Mawioo et al., 2017). The MW radiation of sludge resulted in a prolonged constant rate drying period as a function of the moisture content, an observation that has previously been reported by Bennamoun et al. (2016) and Chen et al. (2014) on the MW drying of sludge. The third drying rate period (i.e., the falling rate drying period) can be observed in Figure 9 and Figure 10 and was most notable when the sludge was irradiated at the higher power outputs above 3 kW. The falling rate drying period was less evident at the lower power outputs of 1 and 1.5 kW. Similar falling rate drying periods during the MW irradiation of sludge have been reported by Bennamoun et al. (2016) and Chen et al. (2014). The authors observed that falling rate drying periods began when the sludge moisture content dropped to 0.7 kg of water dry solids^{-1} min^{-1} and below. This falling rate drying period indicates that the surface of the sludge is no longer completely wet in that the rate at which the water molecules were being transported from the inside of the sludge to the surface of the sludge decreased to below the rate at which the water molecules were being evaporated from the surface of the sludge (Berk, 2018). Therefore, the sludge drying rate decreased at the end of the exposure time. Removing water during the falling rate drying period thus consumes more energy than removing water during the constant rate drying period.

In conventional thermal drying systems involving the use of heated air, steam, or flue gas, the heat transfer mechanism occurs by means of convection and conduction. Thus, the drying process is mostly determined by the water transport mechanisms inside the material to be dried (Léonard et al., 2005; Tao et al., 2005; Li et al., 2016). Léonard et al. (2005) examined the convective thermal drying (hot air at 120°C) of dewatered municipal sludge. The authors reported the presence of a drying rate adaptation period, a constant rate drying period, and a falling rate drying period, yet, the duration of the constant rate drying period was so short that it was indistinguishable during the overall drying process. Tao et al. (2005) and Li et al. (2016) reported similar findings when drying dewatered sludge by means of thermal convective dryers. The authors observed only two phases: the drying rate adaptation period followed by a falling rate drying period. Therefore, it can be inferred that the removal of water during thermal convective drying is governed by an internal water diffusion mechanism (i.e., the mechanism involved in the transport of water molecules from the inside of the material to the surface of the material). External heat needs to penetrate the material (which usually exhibits a low thermal conductivity) to reach the drying front; then, the evaporated water needs to move to the surface of the material through the pores that have been reduced in size due to the same drying effect. Such drying mechanisms are characterised by low drying rates and, thus, long drying exposure times. As reported here and by Bennamoun et al. (2016), MW irradiation extends the constant rate drying period with the falling rate drying period only noticeable at the very end of the drying exposure time.

According to studies by Ni et al. (1999) and Fu et al. (2017) on the MW drying of various materials, the removal of surface water (free water) in the constant rate drying period is attributed to an inverted drying temperature profile with a corresponding increase in the internal pressure gradient. That is, when the material is subjected to MW radiation, the MWs penetrate the material and generate heat from inside the material thereby creating a temperature gradient between the inside of the material (high temperature) and the surrounding environment (low temperature) (i.e., an inverted temperature profile as described by (Shepherd et al., 2018). The high temperature at the core of the irradiated material leads to water evaporation at the core of the material. This, in turn, generates a pressure gradient within the material, driving the water molecules towards the surface of the material (Ni et al., 1999; Fu et al., 2017). Having more of this free water available at the surface of the material results in more free water being evaporated from the surface of the material, suggesting higher sludge drying rates with shorter exposure times and lower energy consumption compared with conventional thermal dryers. These findings are supported by research carried out by Kumar et al. (2016) on the effects of MW irradiation on sliced apples. The authors determined that the vapour pressure was highest in the core of the material and gradually decreased towards the surface of the material. Due to the presence of this vapour pressure gradient, the water molecules can flow from the inside of the material to the surface of the material at a rate approximately 10 times higher than that reported when drying the same material using a thermal convective dryer. Moreover, such penetrative features of the MW radiation and the molecular excitation caused by MWs could contribute to release of water retained both in the complex network of extra cellular polymeric substances and inside of the cell's internal structure and cell membrane (Pino-Jelcic et al., 2006; Yu et al., 2010; Khan et al., 2018; Rao et al., 2019). Therefore, MW radiation increases the amount of surface (free) water available for evaporation thus promoting the occurrence of extended constant rate drying periods, as illustrated in Figure 9 and Figure 10. Noticeably, the MW treatment enhances the rate at which the water leaves the material to be dried, exceeding by far the amount of water that can be transferred to the air by just conventional air drying. In case the air flowrate provided by either an axial ventilator or by any other auxiliary equipment is not large enough (or not provided at all) condensation may occur in the dried material leading to a rewetting of the surface; thus, leading to an increase in the SEC (Fennell et al., 2014; Fennell et al., 2015; Mawioo et al., 2017). Effective removal of water/vapour may be incorporated into such systems by providing a stream of air over the surface of the material under drying as it was the case in this study (Fennell et al., 2014; Fennell et al., 2015). The presence of extended constant rate drying periods has a positive effect on sludge drying since the exposure time (energy) required to remove free water from the material is much lower than the exposure time (energy) required to remove internal water. This unique feature of the MW drying process thus has the potential to provide a remarkable competitive advantage with respect to thermal drying systems given that thermal drying systems require long and energy-intensive exposure times to remove internal water (Ohm et al., 2009; Mujumdar, 2014).

2.3.3 Specific energy output

To evaluate the specific effects of the different MW output powers on the drying rate of the sludge, the sludge drying rates were reported at the same specific energy output (i.e., the energy delivered by the system per mass of sludge sample) as illustrated in Equation 2 for all the evaluated output powers. That is, the sludge drying rates were calculated at a specific energy output (SEO) of 2 MJ kg^{-1} of sludge for all the evaluated output powers (from 1 to 6 kW). The results are described in Figure 11. The sludge drying rates at each evaluated power output were calculated as in the following example. When working at an output power of 1 kW, after an exposure time of 100 minutes (1.21 hours), an energy output of 1.66 kWh (6 MJ) was delivered. The initial mass of the sludge was 3 kg. Dividing the energy output by the initial mass of sludge, a SEO of 2 MJ kg^{-1} of sludge was obtained. At an output power of 1 kW and after 100 minutes of drying, moisture content of 2.4 kg of water kg of dry solids^{-1} (29% DS) and a sludge drying rate of 0.03 kg of water kg of dry solids^{-1} min^{-1} were obtained, as illustrated in Figure 8 and Figure 9, respectively. The same calculations were conducted for all the evaluated MW output powers and the results are presented in Figure 11. The sludge drying rate increased linearly with the increase of the MW output power. Further, the higher the output power, the higher the DS content. Hence, the MW sludge drying process is governed by the rate at which energy is delivered to the sludge rather than by the SEO. The treatment of the sludge at higher MW output powers (at the same energy output) resulted in faster drying (lower exposure time), increasing the throughput potential capacity of the system. The faster drying is a result of both the MW system delivering more energy per unit of time (higher MW output power) and the sludge absorbing more energy per unit of time and unit of volume (higher sludge power absorption density).

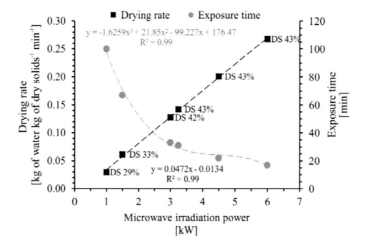

Figure 11: Drying rate and exposure time as a function of the MW output power at a specific energy output of 2 MJ

The sludge power absorption density is described by Equation 8. According to Equation 8, the power absorption density depends on the dielectric properties of the irradiated material, the electric field intensity (i.e., MW power output), the frequency of the MW, and the sample volume. The MW frequency, the initial volume/mass of the sludge, the type of material (sludge), the initial sludge moisture content, and the dielectric properties at the beginning of the experiment were constant throughout the six experimental conditions evaluated. However, the MW output power was increased from 1 to 6 kW with a resulting increase of the electric field intensity, as indicated in the empirical formula described in Equation 9. The increase in the electric field intensity resulted in an increase in the sludge power absorption density and heating rate, as described by Equation 8 and 9. Therefore, increasing the rate at which energy is delivered to the sludge (i.e., increasing the MW output power) resulted in an increase in the capacity of the sludge sample to absorb more energy per unit of time and per unit of volume (i.e., a higher power absorption density), resulting in higher sludge drying rates, as reported in Figure 11. The higher the rate at which the energy is absorbed by the sludge (i.e., the higher the power absorption density), the lower the likelihood that the MW radiation (MW energy) will be reflected towards the dummy water load or absorbed somewhere else in the system. Similar findings have been reported by Robinson et al. (2007) who found a relation between the energy efficiency of the MW drying system and the rate at which the energy was delivered to the irradiated material (biodegradable wastes) rather than the SEO.

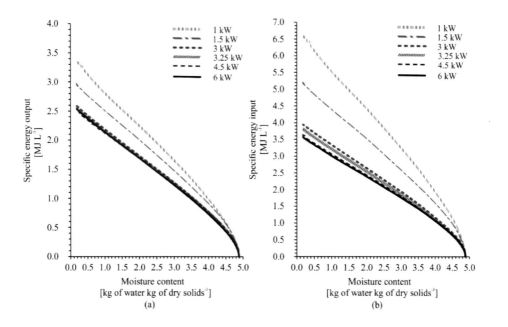

Figure 12: (a) SEO and (b) SEI as a function of the moisture content

Similar findings can be observed when looking at the SEO as a function of the moisture content, as described in Figure 12a. Lower SEO values were reported at the higher MW output powers. In other words, less energy was demanded by the MW system to achieve the same drying results when working at the higher output powers. This lower energy demand could be explained by the higher power absorption density experienced by the sludge at the higher output powers minimising the opportunities for the MW energy to be reflected towards the dummy water load or absorbed somewhere else in the system.

Additionally, similar drying phases as those presented in Figure 10 can be identified when observing the changes in the SEO as a function of the moisture content. A drying rate adaptation period is observed at the right-side of Figure 12a and corresponds to the start of the drying process. The pronounced slope at the beginning of the drying process represents a high SEO per unit of evaporated water. In this drying rate adaptation period, most of the energy was utilised to raise the temperature of the sludge resulting in severely limited water evaporation. After reaching a moisture content of 3.8 kg of water kg of dry solids[-1] (21% DS), a constant rate drying period was observed, as reflected by the consistent slope in Figure 12a. This slope indicates a constant SEO per unit of evaporated water lower than that observed in the drying rate adaptation period. As explained in Section 0, the constant rate drying period is related to the removal of the sludge's free (unbound) water, which requires relatively little energy to be removed from the sludge compared to removal of water in the falling rate period. This low energy demand explains the lower slope in Figure 12a, representing lower energy consumption per unit of evaporated water compared with the drying rate adaptation period. Finally, the falling rate drying period, as described in Figure 10, was not noticeable in Figure 12a. This is likely due to the short duration of this phase, as observed in Figure 10, and the fact that it is only noticeable at evaluated output powers.

2.3.4 MW generation efficiency, energy efficiency, and specific energy consumption

The energy performance of the novel pilot-scale MW system was also evaluated by determining several indicators including MW generation efficiency, energy efficiency, and specific energy consumption.

2.3.4.1 MW generation efficiency

The energy consumed by the novel pilot-scale MW system was measured by a power analyser. The specific energy input (i.e., the energy consumed by the MW system per mass of sludge sample) as a function of the moisture content is presented in Figure 12b. Figure 12a also shows the energy delivered by the MW system as a function of the moisture content. (i.e., the MW power output based on the manufacturer's information) Furthermore, Figure 12b shows the energy consumed by the system (both by the MW power supply and the MW generator) as a function of the moisture content measured using a power network analyser. That is, the ratios between the energy delivered by the MW system (Figure 12a) and the energy consumed by the MW system (Figure 12b) at each of the evaluated powers represent the efficiency of the MW generator (μ_{gen}). These ratios remained constant throughout the drying period and are presented in Figure 13 as MW generation efficiency as a function of the output power. In other words, the efficiency of the electrical energy conversion into electromagnetic energy was evaluated for each output power tested (Equation 5).

For instance, at a MW output power of 1 kW, MW input power of 2 kW was drawn from the power source resulting in MW generation efficiency of 50%. However, as the MW power output increased to 3 kW, MW generation efficiency rose to 70%. Notably, at MW power outputs of above 3 kW, the increase in MW generation efficiency was less pronounced, averaging approximately 70 ± 1.8%. Operating the MW system at the maximum MW output power resulted in the highest MW energy efficiency. Therefore, the optimal performance of the MW generator was obtained when operating the system at the maximum MW power output of 6 kW.

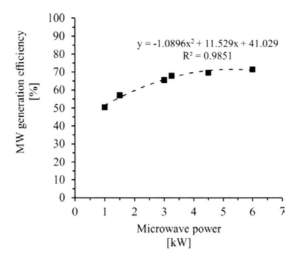

Figure 13: MW generation efficiency as a function of the MW output power

2.3.4.2 Energy efficiency

Energy efficiency (μ_{en}) is represented as the ratio between the theoretical energy demand for the evaporation of the water and the actual energy consumed by the MW unit (Equation 4). Figure 14 shows the energy efficiency for the sludge drying as a function of the moisture content at the evaluated output powers. Overall, energy efficiency increased as the output power increased. Moreover, the same drying phases as described in Section 0 were observed. At the beginning of the drying process (right-side of Figure 14) and up to a moisture content of 3.8 kg of water kg of dry solids^{-1} (i.e., 21% DS), a drying rate adaptation period can be seen. This phase represents a high energy demand per unit of evaporated water since most of the energy was utilised to increase the temperature of the sludge with limited water removal. Thus, the lowest energy efficiency of the evaluated drying process was reported during this period. At the end of the drying rate adaptation period, the energy efficiencies ranged between 36 and 66%for output powers of 1 and 6 kW, respectively. After reaching 3.8 kg of water kg of dry solids^{-1} moisture content, a constant rate drying period was observed. This period was characterised by a constant energy consumption per unit of evaporated water of the drying process. Therefore, a negligible change in energy efficiency was observed in this phase at all the evaluated output powers with energy efficiencies ranging from 33 and 62%found at output powers of 1 and 6 kW, respectively. That is, removing the sludge's free (unbound) water demanded less energy per unit of evaporated water, as reflected in the energy efficiency. The final falling rate drying period, as illustrated in Figure 14, was not noticeable when reporting the energy efficiency as a function of the moisture content.

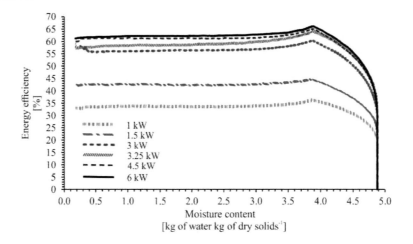

Figure 14: Effect of MW output power on the energy efficiency as a function of sludge moisture content

The energy efficiencies were determined considering the amount of energy drawn from the power source and not the actual energy delivered when irradiating the sludge. Therefore, the energy efficiencies reported in Figure 14 were directly affected by the MW generation efficiency (or inefficiency), as described in Figure 13. Taking a MW generation efficiency of 100%, the maximum energy efficiencies previously reported at the constant rate drying periods would increase to 66 to 89% (from 33 to 62%, respectively) at the evaluated output powers of 1 and 6 kW, respectively. Even at 100% MW energy generation efficiency, a considerable amount of energy is still dissipated in the MW system. Eliminating the energy inefficiencies introduced by MW energy generation contributed significantly to the increase of the energy efficiency of the overall MW system. Nonetheless, there are still other major elements that contribute considerably to energy losses in the MW system.

As discussed in Section 2.3.3, the higher the output power, the higher the rate at which the sludge absorbs the MW energy and thus the lower the amount of MW energy that will be dissipated into the system. The MW energy not absorbed by the irradiated material can be reflected to the MW generator and/or escape the system with the extracted vapour. Therefore, the power absorption density capacity of the material can be directly linked to the MW energy losses (and related energy inefficiencies). Such findings are in agreement with studies carried out on the MW drying of various types of materials (Alibas, 2007; Wang et al., 2009). The authors reported that an increase in the MW power output, and thus in the power absorption per unit of the sample, was followed by an increase in energy efficiency (a reduction in MW energy reflection). Leiker et al. (2004), evaluated the energy efficiency of a MW system for drying wood. The evaluation was carried out by continuously measuring both the MW output power and the reflected power. The authors found that more than 80% of the MW energy was absorbed by the irradiated sample (beech wood), while the remaining 20% of the MW energy was reflected. Thus, energy efficiency as high as 80% was reported (Leiker et al., 2004). These findings are consistent with Swain et al. (2006) who reported a 20% decrease in MW output power after operating a domestic MW appliance continuously for five minutes. In conclusion, the power (energy) supplied by the MW system can be: (i) absorbed by the irradiated material, (ii) returned to the MW generator as reflected power, or (iii) dissipated by being both partially absorbed into the MW cavity and lost with the vapour/condensate (Mawioo et al., 2017). Thus, MW energy losses in the form of reflected power and dissipation into the MW system can explain the energy inefficiencies observed in this research when reporting the MW system energy efficiency (μ_{en}). The higher the output power, the higher the sludge power absorption density, so the lower the chances for MW energy losses.

Moreover, in this research, the MW output powers were not quantified. Rather, these values were based on the manufacturer's ratings and settings. Accordingly, the MW power output could be lower than specified by the manufacturer, leading to an overestimation of the MW output power, MW generation efficiency, and energy efficiency. Jang et al. (2011) reported a drop in MW generation efficiency of 20% when reducing the nominal MW output power from 30 to 10 kW.

2.3.5 Specific energy consumption

The energy performance of the MW unit was also evaluated by determining the specific energy consumption (SEC). The SEC, as described in Equation 1, is defined as the energy consumed by the MW system (measured by the power network analyser at a particular period) to remove 1 L of water from the sludge. The SEC is commonly used to evaluate and compare the energy performance of different types of dryer systems (Kudra, 2012). This parameter (energy) provides an indication of the operational costs of the system. Figure 15a presents the SEC as a function of the sludge moisture content for the entire sludge drying range at the evaluated output powers. At the beginning of the drying process (right-side of Figure 15a and highlighted in Figure 15c), very high SEC values were reported. This corresponds with the very beginning of the drying process during the adaptation drying period when almost no water is evaporated. Consequently, the denominator on the SEC equation (Equation 1) tends to zero, resulting in high SEC values (Figure 15c). The lowest SEC values of approximately 4.5 to 8.5 MJ L^{-1} of water (1.25 to 2.36 kWh L^{-1}) were reported at output powers of 1 to 6 kW, respectively (Figure 15b). As previously discussed, the higher the output power, the higher the energy efficiency and thus the lower the SEC. Moreover, below the moisture content level of 3.8 kg of water kg of dry solids^{-1}, there is a transition from the adaptation drying period to the constant rate drying period. As previously explained, less energy is required during the constant rate drying period demand than the drying rate adaptation period. Therefore, lower SEC values were reported as the drying process progressed (and moisture content reduced).

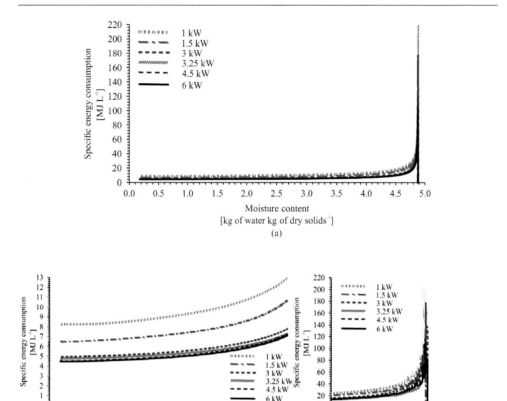

Figure 15: Effect of MW output power on the specific energy consumption as a function of sludge moisture content (a) from 0 to 5 kg water kg dry solid-1, (b) from 0 to 4 kg water kg dry solid-1, and (c) from 4.7 to 5 kg water kg dry solids-1

Figure 16 describes the overall SEC when drying the sludge with an initial moisture content of 4.88 to final moisture content of 0.18 kg of water kg of dry solids^{-1} (i.e., from 17% to 85% DS) as a function of the output power. For instance, a SEC of 8.2 MJ L^{-1} of water (2.3 kWh L^{-1} of water) was obtained when operating the MW system at an output power of 1 kW. Above an output power of 3 kW, much lower SEC values were reported at 4.6 ± 0.16 MJ L^{-1} of water (1.3 ± 0.05 kWh L^{-1} of water). As previously explained, the higher the power output, the higher the energy efficiency of the system and thus the lower the SEC.

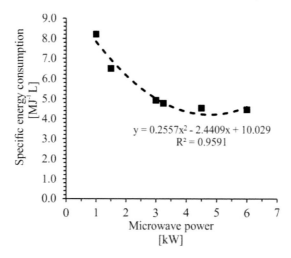

Figure 16: Effect of MW output power on the SEC

Figure 17 shows the SEC for different types of thermal driers (convective and conductive) and MW drying systems. The SEC values reported for the MW drying system evaluated in this study were slightly higher than for thermal driers; however, the SEC values reported in this study were much lower than those reported in the MW pilot-scale study carried out by Mawioo et al. (2017). Therefore, the modifications and innovations introduced in this research mitigated the design difficulties experienced by Mawioo et al. (2017) and can be summarised as follows. Firstly, the selective heating of the irradiated material could be one of the most effective ways of reducing the energy losses associated with the heating of the surface of the cavity and/or other components of the system. The irradiation cavity was constructed from stainless steel, which is characterised by both high conductivity and a high dielectric loss factor and thus does not absorb MW energy (Saltiel et al., 1999; Gupta et al., 2007). Secondly, the provision of a proper vapour extraction system eliminated the in-cavity condensation of the evaporated water during the irradiation process, avoiding the simultaneous evaporation of the water and re-absorption of the condensate. Finally, the provision of a MW generator with a higher quoted MW conversion (generation) efficiency directly improved the overall energy efficiency of the MW system.

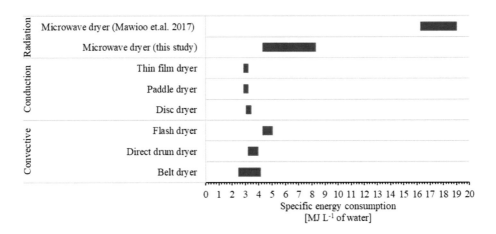

Figure 17: Specific energy consumption (SEC) of different types of dryers (Léonard et al., 2011)

The reported SEC values incorporated the energy expenditure required to heat the sludge to a temperature at which water evaporation would occur. Theoretical energy demand of 0.38 MJ (0.1 kWh) was required to increase the temperature of 1 L of water from room temperature (approximately 10 °C) to 100 °C. The MW drying process produces approximately 0.82 kg vapour per kg of raw sludge[-1]. Such vapour production contains more than enough energy to heat the sludge to the desired temperature. Furthermore, assuming a MW generation efficiency of approximately 60%, heating the sludge with the condensate could represent SEC savings of approximately 0.63 MJ (0.17 kWh) per L of evaporated water, reducing the overall SEC to approximately 3.9 MJ L[-1] (1.1 kWh L[-1]) of water. Additionally, by utilising full-scale industrial MW generators, MW generation efficiencies can be increased by up to approximately 85% (Atuonwu et al., 2019). Therefore, assuming an average MW generation efficiency of 60% in this research, the SEC can be reduced even further to approximately 2.6 MJ L[-1] (0.74 kWh L[-1]). As such, the SEC of the MW drying process is similar to the lower range of SEC found in convective (e.g., belt, direct drum and flash dryer) and conductive (e.g., disc, paddle, thin-film dryers) driers, the energy consumption of which ranges from 2.5 to 5.0 MJ L[-1] of water and 2.9 to 3.4 MJ L[-1] of water, respectively. Moreover, the MW drying process produces a final product (dried sludge) that has a high energetic content (i.e., calorific value) of approximately 18.4 MJ kg[-1] of DS, as indicated in Table 2. The energy content embedded in the final product could produce more energy than the energy required to dry the sludge using MWs.

The results presented and discussed in this study considered the impact of the different MW power outputs on drying performance. As such, it would be desirable to further explore the performance of the MW drying system when other key MW drying features are modified such as the thickness of the sludge in the MW irradiation cavity and the mass to MW output power ratio.

2.4 Conclusions

- The MW sludge drying process extended the duration of the constant rate drying period with consequent high drying rates, short exposure times, lower energy consumption, and high throughput capacity.

- The MW drying performance is governed by the rate at which energy is absorbed by the sludge as heat rather than by the specific energy output. High MW output powers caused high sludge power absorption densities, resulting in increased drying rates and reduced exposure times, increasing the throughput capacity of the system. In other words, less energy was required by the MW system to achieve the same drying results when working at the higher output powers.

- Operating the MW system at the maximum MW output power resulted in the highest MW generation efficiency (i.e., the efficiency of the electrical energy conversion into electromagnetic energy).

- MW energy efficiency increased as the output power increased. Energy losses in the form of reflected power and power that dissipates into the MW system can explain the energy inefficiencies reported. The higher the output power, the higher the sludge power absorption density, so the lower the likelihood of such MW energy losses.

- Similar SEC values were obtained for the MW drying process as for conventional thermal convective and conductive dryers. These results indicate that MW technology is successful at efficiently reducing the sludge moisture content. In addition, MW drying technology may introduce logistic advantages lowering the costs related to sludge transport, handling, and storage.

References

Afolabi OO, Sohail M. (2017). Microwaving human faecal sludge as a viable sanitation technology option for treatment and value recovery–A critical review. Journal of Environmental Management, 187, 401-415.

Akdağ AS, Atak O, Atimtay AT, Sanin FD. (2018). Co-combustion of sewage sludge from different treatment processes and a lignite coal in a laboratory scale combustor. Energy, 158, 417-426.

Alibas I. (2007). Energy Consumption and Colour Characteristics of Nettle Leaves during Microwave, Vacuum and Convective Drying. Biosystems Engineering, 96(4), 495-502.

American Public Health A, Eaton AD, American Water Works A, Water Environment F. (2005). Standard methods for the examination of water and wastewater. Washington, D.C.: APHA-AWWA-WEF.

Atuonwu J, Tassou S. (2019). Energy issues in microwave food processing: A review of developments and the enabling potentials of solid-state power delivery. Critical reviews in food science and nutrition, 59(9), 1392-1407.

Bennamoun L, Chen Z, Afzal MT. (2016). Microwave drying of wastewater sludge: Experimental and modeling study. Drying Technology, 34(2), 235-243.

Berk Z. (2018). Food process engineering and technology: Academic Press.

Bermúdez J, Beneroso D, Rey-Raap N, Arenillas A, Menéndez J. (2015). Energy consumption estimation in the scaling-up of microwave heating processes. Chemical Engineering and Processing: Process Intensification, 95, 1-8.

Bhattacharya M, Basak T. (2016). A review on the susceptor assisted microwave processing of materials. Energy, 97, 306-338.

Bilecka I, Niederberger M. (2010). Microwave chemistry for inorganic nanomaterials synthesis. Nanoscale, 2(8), 1358-1374.

Boldor D, Sanders T, Swartzel K, Farkas B. (2005). A model for temperature and moisture distribution during continuous microwave drying. Journal of food process engineering, 28(1), 68-87.

Chen Z, Afzal MT, Salema AA. (2014). Microwave drying of wastewater sewage sludge. Journal of Clean Energy Technologies, 2(3), 282-286.

Cravotto G, Carnaroglio D. (2017). Microwave chemistry: De Gruyter Berlin, Germany:.

Dominguez A, Menéndez J, Inguanzo MP. (2004). Sewage sludge drying using microwave energy and characterization by IRTF. Afinidad, 61(512), 280-285.

Fennell LP, Boldor D. (2014). Continuous microwave drying of sweet sorghum bagasse biomass. Biomass and Bioenergy, 70, 542-552.

Fennell LP, Bourgeois BL, Boldor D. (2015). Continuous microwave drying of the seeds of the invasive Chinese tallow tree. Biofuels, Bioproducts and Biorefining, 9(3), 293-306.

Fu B, Chen M, Song J. (2017). Investigation on the microwave drying kinetics and pumping phenomenon of lignite spheres. Applied Thermal Engineering, 124, 371-380.

Günther B, Gebauer K, Barkowski R, Rosenthal M, Bues C-T. (2012). Calorific value of selected wood species and wood products. European Journal of Wood and Wood Products, 70(5), 755-757.

Gupta M, Leong EWW. (2007). Microwaves and metals: John Wiley & Sons.

Haque KE. (1999). Microwave energy for mineral treatment processes—a brief review. International Journal of Mineral Processing, 57(1), 1-24.

Hong SM, Park JK, Teeradej N, Lee Y, Cho Y, Park C. (2006). Pretreatment of sludge with microwaves for pathogen destruction and improved anaerobic digestion performance. Water Environment Research, 78(1), 76-83.

Jafari H, Kalantari D, Azadbakht M. (2018). Energy consumption and qualitative evaluation of a continuous band microwave dryer for rice paddy drying. Energy, 142, 647-654.

Jakobsson C. (2014). Sustainable agriculture: Baltic University Press.

Jang S-R, Ryoo H-J, Ahn S-H, Kim J, Rim GH. (2011). Development and optimization of high-voltage power supply system for industrial magnetron. Transactions on Industrial Electronics, 59(3), 1453-1461.

Joardder MU, Karim A, Kumar C. (2013). Effect of temperature distribution on predicting quality of microwave dehydrated food. Journal of mechanical engineering and sciences, 5(December), 562-568.

Joardder MUH, Kumar C, Karim MA. (2017). Multiphase transfer model for intermittent microwave-convective drying of food: Considering shrinkage and pore evolution. International Journal of Multiphase Flow, 95, 101-119.

Kacprzak M, Neczaj E, Fijałkowski K, Grobelak A, Grosser A, Worwag M, Rorat A, Brattebo H, Almås Å, Singh BR. (2017). Sewage sludge disposal strategies for sustainable development. Environmental Research, 156, 39-46.

Kappe CO, Stadler A, Dallinger D. (2012). Microwaves in organic and medicinal chemistry: John Wiley & Sons.

Khan MIH, Nagy SA, Karim MA. (2018). Transport of cellular water during drying: An understanding of cell rupturing mechanism in apple tissue. Food Research International, 105, 772-781.

Kopp J, Dichtl N. (2000). The influence of free water content on sewage sludge dewatering. In Chemical Water and Wastewater Treatment VI (pp. 347-356): Springer.

Kudra T. (2012). Energy performance of convective dryers. Drying Technology, 30(11-12), 1190-1198.

Kumar C, Joardder M, Farrell TW, Karim M. (2016). Multiphase porous media model for intermittent microwave convective drying (IMCD) of food. International Journal of Thermal Sciences, 104, 304-314.

Kumar C, Joardder MUH, Karim A, Millar GJ, Amin Z. (2014a). Temperature Redistribution Modelling During Intermittent Microwave Convective Heating. Procedia Engineering, 90, 544-549.

Kumar C, Karim MA. (2019). Microwave-convective drying of food materials: A critical review. Critical Reviews in Food Science and Nutrition, 59(3), 379-394.

Kumar C, Karim MA, Joardder MUH. (2014b). Intermittent drying of food products: A critical review. Journal of Food Engineering, 121, 48-57.

Lakshmi S, Chakkaravarthi A, Subramanian R, Singh V. (2007). Energy consumption in microwave cooking of rice and its comparison with other domestic appliances. Journal of Food Engineering, 78(2), 715-722.

Leiker M, Adamska M. (2004). Energy efficiency and drying rates during vacuum microwave drying of wood. Holz als Roh-und Werkstoff, 62(3), 203-208.

Léonard A, Blacher S, Marchot P, Pirard J-P, Crine M. (2005). Convective drying of wastewater sludges: Influence of air temperature, superficial velocity, and humidity on the kinetics. Drying technology, 23(8), 1667-1679.

Léonard A, Crine M, Arlabosse P. (2011). Energy efficiency of sludge drying processes.

Léonard A, Vandevenne P, Salmon T, Marchot P, Crine M. (2004). Wastewater Sludge Convective Drying: Influence of Sludge Origin. Environmental Technology, 25(9), 1051-1057.

Li J, Fraikin L, Salmon T, Plougonven E, Toye D, Léonard A. (2016). Convective drying behavior of sawdust-sludge mixtures in a fixed bed. Drying Technology, 34(4), 395-402.

Mawioo PM, Garcia HA, Hooijmans CM, Velkushanova K, Simonič M, Mijatović I, Brdjanovic D. (2017). A pilot-scale microwave technology for sludge sanitization and drying. Science of the Total Environment, 601, 1437-1448.

Mawioo PM, Hooijmans CM, Garcia HA, Brdjanovic D. (2016a). Microwave treatment of faecal sludge from intensively used toilets in the slums of Nairobi, Kenya. Journal of Environmental Management, 184, 575-584.

Mawioo PM, Rweyemamu A, Garcia HA, Hooijmans CM, Brdjanovic D. (2016b). Evaluation of a microwave based reactor for the treatment of blackwater sludge. Science of the Total Environment, 548, 72-81.

Mishra RR, Sharma AK. (2016). Microwave–material interaction phenomena: heating mechanisms, challenges and opportunities in material processing. Composites Part A: Applied Science and Manufacturing, 81, 78-97.

Mujumdar AS. (2014). Handbook of industrial drying (4th ed.): CRC press.

Ni H, Datta A, Torrance K. (1999). Moisture transport in intensive microwave heating of biomaterials: a multiphase porous media model. International Journal of Heat and Mass Transfer, 42(8), 1501-1512.

Ohm T-I, Chae J-S, Kim J-E, Kim H-k, Moon S-H. (2009). A study on the dewatering of industrial waste sludge by fry-drying technology. Journal of hazardous materials, 168(1), 445-450.

Pino-Jelcic SA, Hong SM, Park JK. (2006). Enhanced anaerobic biodegradability and inactivation of fecal coliforms and Salmonella spp. in wastewater sludge by using microwaves. Water environment research, 78(2), 209-216.

Pitchai K, Birla SL, Subbiah J, Jones D, Thippareddi H. (2012). Coupled electromagnetic and heat transfer model for microwave heating in domestic ovens. Journal of Food Engineering, 112(1), 100-111.

Raheem A, Sikarwar VS, He J, Dastyar W, Dionysiou DD, Wang W, Zhao M. (2018). Opportunities and challenges in sustainable treatment and resource reuse of sewage sludge: a review. Chemical Engineering Journal, 337, 616-641.

Rao B, Su X, Lu X, Wan Y, Huang G, Zhang Y, Xu P, Qiu S, Zhang J. (2019). Ultrahigh pressure filtration dewatering of municipal sludge based on microwave pretreatment. Journal of Environmental Management, 247, 588-595.

Robinson JP, Kingman SW, Snape CE, Shang H. (2007). *Pyrolysis of biodegradable wastes using microwaves.* Paper presented at the Proceedings of the Institution of Civil Engineers-Waste and Resource Management.

Saltiel C, Datta AK. (1999). Heat and mass transfer in microwave processing. Advances in heat transfer, 33(1), 1-94.

Schaum C, Lux J. (2010). Sewage sludge dewatering and drying. ReSource–Abfall, Rohstoff, Energie, 1, 727-737.

Shepherd B, Ryan J, Adam M, Vallejo DB, Castaño P, Kostas E, Robinson J. (2018). Microwave pyrolysis of biomass within a liquid medium. Journal of analytical and applied pyrolysis, 134, 381-388.

Soltysiak M, Erle U, Celuch M. (2008). *Load curve estimation for microwave ovens: experiments and electromagnetic modelling.* Paper presented at the MIKON 2008-17th International Conference on Microwaves, Radar and Wireless Communications.

Stuerga D. (2006). Microwave-material interactions and dielectric properties, key ingredients for mastery of chemical microwave processes (Vol. 2): WILEY-VCH Verlag GmbH & Co. KGaA.

Swain MJ, Ferron S, Coelho AIP, Swain MVL. (2006). Effect of continuous (intermittent) use on the power output of domestic microwave ovens. International journal of food science & technology, 41(6), 652-656.

Tao T, Peng X, Lee D. (2005). Structure of crack in thermally dried sludge cake. Drying technology, 23(7), 1555-1568.

Vadivambal R, Jayas D. (2010). Non-uniform temperature distribution during microwave heating of food materials—A review. Food and bioprocess technology, 3(2), 161-171.

Vesilind PA. (1994). The role of water in sludge dewatering. Water Environment Research, 66(1), 4-11.

Wang Z-F, Fang S-Z, Hu X-S. (2009). Effective diffusivities and energy consumption of whole fruit Chinese jujube (Zizyphus jujuba Miller) in microwave drying. Drying technology, 27(10), 1097-1104.

Welsh Z, Kumar C, Karim A. (2017). Preliminary Investigation of the Flow Distribution in an Innovative Intermittent Convective microwave Dryer (IMCD). Energy Procedia, 110, 465-470.

Yu Q, Lei H, Li Z, Li H, Chen K, Zhang X, Liang R. (2010). Physical and chemical properties of waste-activated sludge after microwave treatment. Water Research, 44(9), 2841-2849

3

Microwave treatment of municipal sludge

Effects of the sludge thickness and sludge mass load on the microwave drying performance

This chapter is based on: Kocbek E, Garcia HA, Hooijmans CM, Mijatović I, Brdjanovic D. Microwave treatment of municipal sewage sludge: Effects of the sludge thickness and sludge mass load on the drying performance. Submitted to Journal of Environmental Management, 2021.

Abstract

Sludge transportation costs can represent a big percentage of the total sludge treatment costs generated at a municipal wastewater treatment plant (WWTP). These expenses may be mitigated by reducing the sludge volume by using thermal drying approaches. This study investigated a pilot-scale microwave (MW) system as a potential alternative approach for drying sewage sludge. In the present study, the influence of the initial sludge mass (1, 1.5, 3 and 4.5 kg) and sample thickness (45, 84, 105 and 150 mm) at constant MW power output on the drying rates, energy efficiencies, and specific energy consumption of sludge were assessed. The results showed that the rate at which the MW energy is absorbed per unit of volume (i.e., power density) has a direct impact on the MW system's throughput capacity. As such, it is possible to enhance the performance of the MW system by modifying the power density across the sample by adjusting the power-to-mass ratio; specifically, the use of a lower applied mass during treatment can result in reduced exposure time and increasing the drying rates. However, this could lead to an increase in the reflective power energy losses of the system, which may undermine the system's energy performance. The condition deteriorates when the thickness of the sludge layer that is exposed to radiation is significantly greater than the penetration depth of the MW irradiation. The larger the sample thickens, the higher the degree of non-uniform distribution of the electromagnetic energy within the material; thereby, decreasing the energy efficiency. The limitations associated with the MW penetration depth were to a certain extent counteracted by the reduction in the moisture content of the sludge while drying, lowering the dielectric properties of the sludge. The outcomes of this study demonstrate that the uniform distribution of the MW energy that the material absorbs over a period of time represents a fundamental element in the improvement of the system's energy performance. To achieve relatively effective heating uniformity with the least amount of energy loss, there is a requirement to determine the optimum MW parameters and thickness and mass of the sample.

3.1 Introduction

The rapid global population growth has raised concerns regarding the proper management and disposal of municipal sewage and non-sewage (septic tank sludge) sludge. In Europe, over 10 million tons of dry solid (DS) are produced annually with an specific sludge annual production between 0.1 and 30.8 kg per population equivalent (p.e.) (Comission, 2008; Kelessidis et al., 2012). The reported sludge production values were expressed as dry solids; however, the sludge contains significant amounts of water. For instance, the sludge moisture content at a wastewater treatment plant (WWTP) after mechanical dewatering processes, e.g. centrifugation or filtering, could be of approximately 80% (20% DS) (Flaga, 2005). Correspondingly, the previously mentioned value of 10 million tons of DS produced annually in Europe can be converted into 50 million tons of actual (wet) sludge produced annually in Europe requiring proper treatment and disposal.

Generally, sludge contains substantial amounts of phosphorous (P), nitrogen (N), various micro nutrients, and organic matter. In many regards these significant components make the sludge a good source as a soil enhancer and a cheap fertilizer. Sewage sludge has been utilized via direct land application for agricultural amendment (Fytili et al., 2008). However, xenobiotic and emerging organic pollutants (e.g. micro plastics, hormones, and antibiotics etc.) can be present in significant amount in municipal sludge along with pathogens and heavy metals (Köhler et al., 2012; Herzel et al., 2016). Because of the emerging risks associated with direct application of sewage sludge on land, many countries have banned those practices (or are in the process of doing so); thus, other sludge treatment/disposal alternatives such as (co-)incineration have been gaining traction (Christodoulou et al., 2016) despite their higher treatment costs compared to conventional sludge treatment/disposal options (Jakobsson, 2014). One of the most influencing factors determining the sludge management/treatment costs is the transportation costs from the point of generation to the ultimate treatment/disposal location; such costs may account to over half of the total sludge management costs. Europe has a large number of medium and small WWTPs, with a design treatment capacity of to cover the wastewater generated by approximately 10,000 population equivalents (PE). For example, Slovenia has approximately 250 municipal WWTPs in operation; however, only 10% of such facilities have a treatment capacity larger than 10,000 PE (LeBlanc et al., 2009). Similar situations are observed in other European countries as follows. There are approximately 18,000 WWTPs in France, and approximately 93% of them serve small communities of 10,000 PE or less (Ledakowicz et al., 2019). For such small WWTPs it is not economically feasible to implement *in-situ* advanced sludge treatment processes to locally deal with the sludge and in such way to eliminate (or considerably reduce) the final sludge disposal needs (for instance by implementing an in-situ sludge incineration plants). So, in such circumstances the sludge needs to be transported to centralized treatment/disposal sites requiring high sludge transportation costs. As a result, different approaches have been implemented in place, at the WWTP, to reduce the mass and volume of the sludge (Schaum et al., 2010); thus, alleviating the sludge transportation costs. Such processes for sludge dewatering and drying usually include a gravity thickener pre-treatment dewatering step; on subsequent processes, the sludge is further dewatered and/or dried. Mechanical dewatering usually involves the use of belt presses, centrifuges, and similar processes. Mechanical dewatering can achieve a sludge moisture content of approximately 70% (30% DS). Sludge drying can further reduce the sludge moisture content below 10% (90% DS). Sludge dryer alternatives include belt dryers, rotary dryers, drum dryers, paddle, disc dryers, and the recently introduced microwave (MW) dryer systems.

MW systems have been proposed as a promising drying alternative offering many advantages over the conventional convective and conductive drying systems including: high drying rates, contactless heating, and instantaneous heating, among others (Bermúdez et al., 2015; Mawioo et al., 2017). The MW energy causes dipolar molecules to rotate within the irradiated material causing friction which results in heat generation within the material (Bhattacharya et al., 2016). The MW energy penetrates the irradiated material generating heat from the inside of the material (Ni et al., 1999; Kumar et al., 2016a). Therefore, this effect produces a temperature gradient from the inside to the outside of the material (i.e., exhibiting high-temperatures in the material interior and low-temperatures in the external environment). Consequently, water is easily converted to vapour in the interior of the material generating a pressure gradient propelling the water molecules to the surface of the irradiated material (Ni et al., 1999; Kumar et al., 2016a). Once the water molecules reach the surface of the material, there are more chances for that free water to leave the material (evaporate). So, the availability of such free water at the surface of the material means that a larger amount of water is evaporated from the surface resulting in larger drying rates (lower drying exposure times) compared to conventional (convective and conductive) thermal driers. Such drying mechanisms have not been observed in conventional drying systems.

In conventional (convective and conductive) drying systems, the heat transfer proceeds from an external source of heating raising the temperature of the surface of the material being heated by conduction or convection mechanisms (Stuchly et al., 1972). Thus, the drying limiting step for the conventional heating is the removal of the internal water molecules from the interior of the irradiated material; removing such internal water consumes approximately two-thirds of the total drying time (Mujumdar et al., 2000). In the MW drying system, due to the generated inverse temperature gradient, the drying rates of the irradiated material can be increased up to approximately 10 times (Dominguez et al., 2004; Guilong et al., 2010). Moreover, it allows a more precise control of the heating process and heating conditions reducing the risks of surface overheating. However, not all the materials can be dried by the use of MW radiation, since not all the materials can absorb MW energy (Mello et al., 2014). The aptitude of a material to absorb and convert the MW energy into heat is governed by the material dielectric properties, namely its dissipation factor (tan δ). The tan δ is defined as the ratio of the dielectric loss factor (ε'') to the dielectric constant of the material (ε') (Mawioo et al., 2017). The amount of MW energy that can be absorbed by the material is represented by dielectric loss factor, whereas the ability of a material to convert the energy into heat is represented by dielectric constant. Therefore, the higher the dissipation factor, the higher the aptitude for a material to convert MW energy into heat (Haque, 1999). Sludge is a good example of a material suitable to convert the MW energy into heat.

As revealed by previous studies, MW treatment has been effective in drying and reducing the pathogen content of the sludge. Menendez et al. (2002) successfully dried anaerobic sewage sludge up to a final moisture content of 30% (70% DS). Similar observation were reported for fresh faecal sludge Mawioo et al. (2016a), blackwater sludge (Mawioo et al., 2016b), septic tank sludge (Mawioo et al., 2017), and dewatered waste activated sludge (WAS) (Kocbek et al., 2020). The authors concluded that the efficiency of the MW drying system was primarily determined by the rate of absorption and conversion of the MW energy into heat. The higher the MW output power, the higher the MW energy absorption density; thereby, increasing the drying rates (reducing the exposure times needed to achieve a certain level of sludge dryness). Most of the studies in the literature have focused mainly on assessing the moisture removal of the sludge as a function of either the amount of sludge being irradiated, or the applied MW power output. However, the effects of the sludge thickness on the drying performance of the MW technology have not been deeply evaluated so far.

Assessing the impact of the sludge thickness on the drying performance is especially relevant. Evenly heating the entire volume of the MW irradiated material (volumetric heating), among other factors, depends on the capacity of MW energy to penetrate the entire depth of the irradiated material which can strongly depend on MW frequencies and dielectric properties of the sludge. Antunes et al. (2017), when evaluating the dielectric properties of municipal sludge, reported a narrow MW penetration depth (at a frequency of 2,450 MHz) of less than 5 and 14 mm at a sludge moisture content of 80 and 5% (20 and 95% DS), respectively. Thostenson et al. (1999) reported that when the thickness of the material is greater than the MW penetration depth, only the surface of the material is heated by the direct action of the MW energy, while the rest of the material is heated through conduction mechanisms. That is, beyond the material penetration depth, the volumetric heating of the sludge due to the MW irradiation cannot be achieved. Thus, increasing the sludge thickness beyond the MW penetration depth would have a negative effect on the uniform distribution of the electromagnetic energy throughout the material. As such, this would also impact on the temperature distribution within the irradiated material, and on the overall MW drying efficiency. Other authors also agreed with the findings reported by Thostenson et al. (1999); (Darvishi et al., 2018; Azimi-Nejadian et al., 2019; Çelen, 2019). However, Jafari et al. (2018) when MW drying rice paddy at a power output of 450 W, reported that an increase in the thickness layer from 6 to 18 mm at constant mass increased the energy efficiency of the MW system; the specific energy consumption (the energy consumed for evaporating one litre of water) decrease from 53 MJ L^{-1} (at 6 mm) to 24 MJ L^{-1} (at 18 mm) (Jafari et al., 2018). According to the authors, increasing the thickness of the material increases the MW energy absorption by the irradiated sample (reduced MW energy reflection). Undoubtedly, the thickness of the irradiated material introduces a strong impact on the overall drying performance of a MW drying system. To our knowledge, our study is the first to evaluate the effects of the sludge thickness on the overall MW drying performance (all other variables being equal such as the initial mass to MW power output ratio).

The main objective of this research was to evaluate the effects of the sludge thickness on the drying and energy performance of a pilot-scale MW system. Municipal sludge MW irradiated at different sludge thickness layers ranging from 45 up to 180 mm at a constant initial mass to

MW power output ratio. Additional evaluations were also conducted assessing the effects of different loads of sludge, at a constant thickness, on the drying and energy performance of the pilot-scale MW system. The exposure times, drying rates, energy efficiencies, and specific energy consumptions were determined at the evaluated conditions.

3.2 Materials and methods

3.2.1 Experimental design

The effect of the sludge thickness on the MW drying performance was assessed by taking several dewatered (centrifuged) WAS sludge samples taken at a local municipal WWTP in Slovenia. The samples were further irradiated in a pilot-scale MW system at a defined MW power output. The samples were loaded in the MW system at a defined thickness and sample mass and the performance of the MW system at each sludge thickness level and sample mass was evaluated by determining the drying rates, energy efficiencies and the specific energy consumption as described below.

3.2.2 Experimental pilot-scale MW system

A pilot MW system that was designed to treat a variety of sludge types that contained various solid and water content; for instance, septic tank sludge, fresh faecal sludge, and sewage sludge. Figure 18 presents a comprehensive schematic image of the MW system. The system incorporated a stainless-steel cylindrical MW irradiation cavity that housed a polypropylene (PP) oval vessel. The holding vessel was placed on a rotating PP disc that was powered by an electromotor to rotate at a speed of 1 rpm. Rotating this sludge in this way served to ensure uniform irradiation. The supplementary equipment consisted of a ventilation unit, which extracted any condensate; an air filtration system, which controlled odour; a power supply; and an MW magnetron, which operated at a 2450 MHz frequency and a 3 kW MW output power and delivered the required power via a standard rectangular waveguide WR340 (86.36 × 53.18 mm) that was linked to a circulator. The MW power passed to the MW cavity via a waveguide with an isolator that was connected to the MW head. This aspect of the system was designed such that the MW power could flow in a forward manner but not in reverse. Reflected power was absorbed by a dummy load that was linked with the waveguide circulator. Any elements that had the potential to damage the MW magnetron, such as dust and humidity, were managed by incorporating a teflon window between the MW chamber's inlet and the MW generator's outlet. A magnetron-cooling water-based structure was integrated to prevent overheating in the power supply and the MW magnetron. Demineralised water was supplied at a 600 L h^{-1} flowrate. Three distinct filler types were employed to selectively increase the air filtration system's adsorption. These fillers consisted of activated carbon soaked in sodium hydroxide (NaOH), activated carbon soaked in phosphoric acid (H3PO4), and aluminium oxide (Al2O3) with potassium permanganate (KMnO4). Any variations in the mass of the sludge that occurred during the process of drying were monitored using a single point load cell (Mettler Toledo) that had an error below 0.016 g and a 5 g resolution. The electrical energy that was provided to the MW unit was also measured on a continual basis through the use of a power network analyser (Etimeter). Three fibre optic temperature probes

OPTOcon GmbH) with 1°C precision were placed along the centre of the sample holding vessel (i.e., sludge sample). These probes were employed to determine the temperature inside the sludge sample at the different evaluated thickness levels. The temperature measurements were carried out, in absence of rotation movement caused by the turntable (motor).

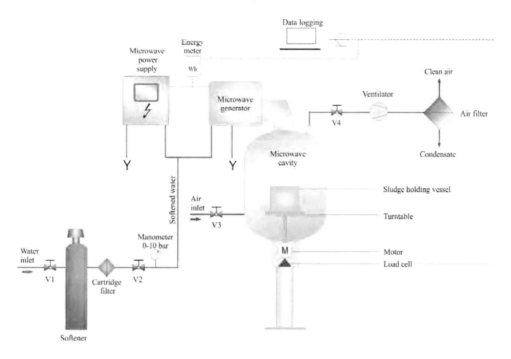

Figure 18: Schematic representation of the experimental pilot-scale MW system (Kocbek et al., 2020)

3.2.3 Analytical determinations

3.2.3.1 Total solids (TS), and absolute density determination

Per the approach described by American Public Health et al. (2005), the TS were measured according to the SM-2540D gravimetric methods. The DS% was the equivalent of the TS concentration denoted in percentage form; the moisture content was computed by subtracting the DS% from 100. The sludge's absolute density was measured in a helium atmosphere through the use of an AccuPyc 1330 automatic gas pycnometer (Micromeritics Inc., USA). Both the moisture content and absolute density values were determined previously by the same author (Kocbek et al., 2020), who reported an average value of 4.88 kg of water kg of dry solids^{-1} (i.e., 15% DS) and 1.4 g cm^{-3} (Chapter 4).

3.2.4 Experimental procedure

The WAS samples were collected from the wastage of an activated sludge process at the WWTP Ptuj, Slovenia. The sludge was mechanically dewatered (centrifuged with the addition of a polymer). The sludge samples were collected on the same day before conducting the MW evaluations. The samples were weighed and placed in the PP cylindrical-shaped vessels within the MW cavity on a revolving turntable. The experimental conditions for the evaluated samples including the MW power output, the sludge mass, the exposed surface area, and sample thickness are described in Table 3. The sample thickness values are indicative of the sludge thickness levels (height) determined according to the formula given for volume of the cylinder. The cylinder volume is defined as $\pi r^2 h$, where r is the radius of the circular end of the PP cylindrical shaped vessel and h is the sludge height (thickness). The volume of the sludge was determined by measuring the sludge mass and sludge density. From the known sludge volume and circumference of the PP vessel, the sludge thickness was calculated. Two different set of experiments can be distinguished. On the first set of the experiments, (experiments 1 to 4 as in Table 3), the mass load was increased from 1 to 4.5 kg at the same MW power output and sample thickness of 3 kW and approximately 49 ± 3 mm, respectively. The changes in the sample mass load at constant thickness were achieved by varying the surface area of the sludge. On the second set of experiments (experiments 5 to 8 as in Table 3), the thickness of the sludge was increased from 46 to 150 mm at the same MW power output and sludge mass of 3 kW and 3 kg, respectively. The surface area of the sludge was also modified to achieve the target thickness level. The samples were MW dried until reaching a sludge moisture content of 0.18 kg water kg dry solids[-1] (85% DS). During the evaluations, change in sludge mass, temperature, energy consumption and exposure time were monitored as shown in Table 3. All the evaluations were performed either in duplicates, or triplicates; the average results were reported.

Table 3: Experimental design

| Experiment | Experimental parameters and variables | | | | Monitoring parameters | | | |
| | Microwave power | Mass | Exposed surface area | Thickness | Mass | Temperature | Energy consumption | Exposure time |
	[kW]	[kg]	[m^2]	[mm]	[kg]	[°C]	[kWh]	[min]
1	3.0	1.0	0.019	50.0	•		•	•
2	3.0	1.5	0.027	52.0	•		•	•
3	3.0	3.0	0.062	45.0	•		•	•
4	3.0	4.5	0.086	49.0	•		•	•
5	3.0	3.0	0.019	45.0	•	•	•	•
6	3.0	3.0	0.027	84.0	•		•	•
7	3.0	3.0	0.033	105.0	•		•	•
8	3.0	3.0	0.062	150.0	•		•	•

3.2.5 Data analysis

3.2.5.1 Specific energy consumption (SEC)

The SEC was calculated as shown in Equation 10:

$$SEC = \frac{P_{in;elect} \cdot t}{m_{eva}} \tag{10}$$

where the SEC is the specific energy consumption [kJ L^{-1} of water], $P_{in;elect}$ is the input power consumed by the system during the drying process [kW], t is the exposure time [s], and m_{eva} is the amount of evaporated water at a specific exposure time [L] . The $P_{in;elect}$ was measured using a power network analyser (Etimeter), as described in Section 3.2.2.

3.2.5.2 Energy efficiency (μ$_{en}$)

The μ$_{en}$ is the ratio between the theoretical energy demand for evaporating the water and the energy consumed by the MW unit during the drying process, the calculation for which is shown in Equation 11 (Jafari et al., 2018):

$$\mu_{en} = \frac{(m_{sample} \cdot c_p \cdot \Delta T) + (m_{eva} \cdot h_{fg})}{P_{in;elect} \cdot t} \cdot 100 \tag{11}$$

where μ$_{en}$ is the energy efficiency [%], m_{sample} is the initial mass of sludge (kg), c_p is the specific heat capacity of the water [kJ kg^{-1} °C^{-1}], ΔT is the temperature difference of the sample between the exposure time t and the start of the treatment, and h_{fg} is the latent heat of the water [kJ kg^{-1}] (2,257 kJ kg^{-1} at 100 °C as reported by Haque (1999)). The influence of sensible heat on total energy efficiency was considered assuming the sludge sample reached a temperature of 100 °C before water evaporation took place.

3.2.5.3 Sludge moisture content (X)

The X was calculated as shown in Equation 12 (Chen et al., 2014):

$$X = \frac{m_t - m_d}{m_d} \tag{12}$$

where X is the moisture content of the sludge [kg of water per kg of dry solid^{-1}], m_t is the sample mass at time t, m_d is the total mass of dry solids in the sample [kg]. The total mass of dry solids in the samples was determined as the DS of the raw C-WAS sludge sample, as described in Section 3.2.3.1. The m_t was continuously determined by the point load cell located in the MW irradiation cavity, as described in Section 3.2.2. Therefore, X was continuously measured.

3.2.5.4 Drying rate (D_R)

The D_R was calculated as shown in Equation 13 (Chen et al., 2014):

$$D_R = \frac{dX}{dt} \tag{13}$$

where D_R is the drying rate [kg of water per kg of dry solid^{-1} min^{-1}] and X is the moisture content of the sludge at a specific exposure time.

3.2.5.5 Electromagnetic heat generation

Lambert's law and Maxwell's equations are commonly employed for quantifying the electromagnetic energy generation within the irradiated material. Maxwell's equations are traditionally used to compute the power distribution for thin materials, whereas Lambert's law is used to compute the power distribution for thick materials (Liu et al., 2005b; Budd et al., 2011; Khan et al., 2020). According to Maxwell equation, the power absorption density (i.e., the amount of power absorbed by a material per unit of volume) (P_d) is proportional to the input power, which relates to the electric field intensity (E), as shown in Equation 14 (Stuerga, 2006; Gupta et al., 2007):

$$P_d = 2\pi f \varepsilon_0 \varepsilon' |E|^2 \tag{14}$$

where P_d is the amount of absorbed power per unit volume [Wm^{-3}], f is the microwave frequency [s^{-1}], ε_0 is the permittivity of free space [8.85×10^{-12} Fm^{-1}] and E is the electric field intensity [Vm^{-1}]. The electric field intensity can be calculated as described in Equation 15 (Soltysiak et al., 2008; Pitchai et al., 2012).

$$E = \sqrt{\frac{2P_{out,micr}}{1 - |S_{11}|^2}} \tag{15}$$

where S_{11} is the reflection coefficient associated with the fraction of the power reflected by the sample. According to Lambert's law, the MW energy (power) absorption in a thick product, is expressed via the following relationship (Chandrasekaran et al., 2012)

$$P(x) = P_0 e^{(-2\alpha x)} \tag{16}$$

where, P(x) is the power dissipated at a distance x [W], P_0 is the incident power at the surface of the sample [W], and α is the attenuation constant [m^{-1}], which is a function of wavelength of radiation (λ) [m], dielectric constant and loss tangent. The attenuation constant is given by the following equation (Chandrasekaran et al., 2012):

$$\alpha = \frac{2\pi}{\lambda} \sqrt{\frac{\varepsilon'((1 + tan^1\delta)^{\frac{1}{2}} - 1)}{2}} \tag{17}$$

3.2.5.6 Penetration depth (d_p)

The MW penetration depth is defined as follows:

$$d_p = \frac{c}{2\sqrt{2}\pi f}\left\{\varepsilon'[\sqrt{1 + tan\delta^2} - 1]^{-1/2}\right\} \tag{18}$$

where dp represents the penetration depth [m] and c the velocity of light [3×10^8 ms^{-1}].

3.3 Results and discussion

3.3.1 MW sludge drying performance: drying rates

This section discusses the effects of the sludge thickness and sample mass on the performance of the MW drying system focusing on the sludge drying rates; main parameters discussed in this section include the exposure time (t) for achieving a certain degree of dryness and the drying rates (D_R).

3.3.1.1 Effects of the sludge mass on the drying performance

Sólyom et al. (2011) reported that the MW drying is affected by the initial mass of sludge exposed to the MW irradiation reflecting the amount of MW energy absorbed by the sludge sample. Four different sludge samples at 1, 1.5, 3 and 4.5 kg were MW dried at a constant thickness and MW power output of 49 mm and 3kW, respectively (experiments 1 to 4 in Table 3) to evaluate the effect of the initial sludge mass on the MW drying performance. Figure 19a shows the reduction of the sludge moisture content as a function of the exposure time required to treat the irradiated material from its initial moisture content of 4.88 kg water kg dry solids^{-1} (15% DS) to a final moisture content of 0.18 kg water kg dry solids^{-1} (i.e., 85% DS content). For all the evaluated samples the sludge moisture content decreased with the exposure time. The exposure times decreased as the initial sludge mass of the sample decreased (i.e., approximately 15 and 70 minutes were needed to dry 1 and 4.5 kg of initial sludge samples, respectively up to a final 85% DS concentration). In addition, higher rates were observed for the lower initial sludge mass samples compared to the higher ones as observed in Figure 19b. That is, not only lower exposure times were needed for the lighter sludge samples, but also the rate at which the water was removed was much higher. This could be explained considering the MW power applied to the sludge per unit mass or volume (i.e., the power density as described in Equation 14. The MW power output was the same for all the evaluated samples, so the power density increased with an increase in the initial power output to mass ratio; thus, leading to a faster removal of water contained in the sludge.

A similar finding was reported in the literature (Chen et al. (2014); Bennamoun et al. (2016); Mawioo et al. (2016a); Mawioo et al. (2016b). The performance of the MW-based drying system could be optimized by simply adjusting the power density across the sample; treating sludge at a higher power density (high power output to mass ratios) can lead to higher drying rates and shorter exposure time. The fundamental processes supporting such explanation was previously reported in detail by the authors in Kocbek et al. (2020).

Figure 19b (Krischer's curve) indicates three distinctive drying periods commonly observed in MW sludge drying processes as follows (Chen et al., 2014; Bennamoun et al., 2016; Mawioo et al., 2016b): (i) a drying rate adaptation period, (ii) a constant rate drying period, and (iii) a falling rate drying period. The sludge drying rates raised rapidly at the start of the drying, followed by a constant drying rate. Finally, a drop in the drying rates were observed at the end of the drying process. Such drying periods describe changes in the heating and evaporation of the water molecules present in the sludge. For instance, the steep rise in the drying rate during the adaptation period can be associated with the presence of a large amount of water molecules starting to interact with the MW electromagnetic radiation. Heat is generated when these molecules interact with MW radiation resulting in an increase on temperature and drying rates. Following such an adaptation drying phase, the water began to be removed from the sludge at a steady-state condition (constant rate drying periods). The MW energy converted into heat was used in that phase for the vaporization of mostly unbound water located at the surface of the sludge. The evaporated water at the surface of the sludge was continuously replaced by water coming from the interior of the sludge (Flaga, 2005; Mawioo et al., 2016a). The water transport from the interior to the surface of the sludge has been well reported in the literature, and it can be mostly attributed to the pressure gradient generated during MW drying due to the observed inverted temperature profile in the sludge (Kumar et al., 2016b). When the water flow from the interior to the surface of the irradiated material could not cope with the rate at which the water is evaporated from the surface of the sludge, the drying rate decreased and the process entered into the falling rate periods. A more detailed description of the drying rate periods was reported by Keey (1991) and Bennamoun et al. (2013), and also reported by the authors of this study, in Kocbek et al. (2020).

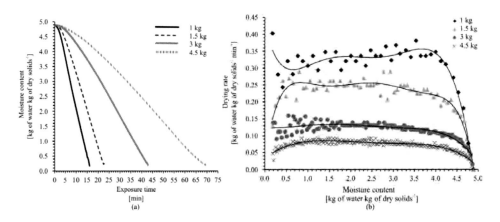

Figure 19: Effect of the initial mass of sludge on (a) the reduction of the sludge moisture content as a function of the exposure time, and (b) the drying rates as a function of the sludge moisture content

3.3.1.2 Effects of the sludge thickness on the drying performance

Four additional experiments were carried at different sludge thickness from 45 to 150 mm (experiments 5 to 8 in Table 3) at a constant initial mass of sample and MW power output of 3 kg and 3 kW, respectively to evaluate the effect of the sludge thickness on the MW drying performance. Figure 20a shows the variation in the sludge moisture content as a function of exposure time for all the evaluated samples at the different thicknesses. The exposure time required to reach the target sludge moisture content of 15% DS increased with increasing the thickness layer. In addition, as shown in Figure 20b the drying rates seemed to slightly decrease

with increasing the sludge layer thickness. Therefore, these results indicate that the drying performance was negatively affected by the sludge layer thickness; as the thickness of the sludge increased, lower drying rates and larger exposure time were needed to achieve the drying target of the sludge. Such an effect could be attributed to the narrow MW penetration depth within the sludge samples, defined in Equation 18, that has been reported to vary between 5 and 14 mm, depending on sludge moisture content (Antunes et al., 2018). In addition, the loss of the MW energetic power could occur as the MW energy travels through the irradiated material. This effect is known as Lambert's law and states that the MW power decay exponentially into the material (Liu et al., 2005a).

Such an effect can be observed by monitoring the temperature within the irradiated sludge. Figure 21a shows the temperature profile measured by fibre optic sensor at various thickness levels, $\delta_1 = 5$ mm, $\delta_2 = 25$ mm, and $\delta_3 = 40$ mm in sample mass load of 3 kg, irradiated at MW power output of 3 kW. The result presented in Figure 21, corresponds to experiment nr. 5 described in Table 3. The position of the fibre optic sensors within the sludge were located as described in Figure 21. At the point δ_1, near the surface of the sludge, the temperature increased to the water boiling point ($T_{\delta_1}=100°C$) within two minutes of the exposure to the MW irradiation. At the other two evaluated points (δ_2 and δ_3) the temperature increased gradually up to a final temperature of 100°C within 8.5 to 13 minutes. These results indicate that the penetration depth relative to the total sample layer thickness (45 mm) has a substantial effect on the temperature distribution inside the material; the thicker the sample, the longer the exposure time to reach the water boiling temperature.

Such non-uniform heating behaviour observed for the MW irradiated sludge can be up to a certain extent counteracted by the levelling effect caused by the removal of the moisture content of the sludge (Zhu et al., 2015; Gaukel et al., 2017). During the MW drying of the sludge, the changes in the sludge temperature and the water removal caused variations on the sludge dielectric properties (Antunes et al., 2018). On sludge samples exhibiting high moisture content, that high water content yields excellent dielectric properties to the sludge; thus, such high moisture content sludge would very efficiently absorb and convert MW energy into heat (Antunes et al., 2018). Therefore, the MW radiation effectively heats the sludge regions exhibiting a high moisture content. The temperature changes within the sludge are usually followed by a reduction in the sludge moisture content; therefore, resulting in a change (decrease) on the dielectric properties of the sludge. Such decrease on the dielectric properties means that the MW radiation could better penetrate the irradiated material (i.e., as the dielectric properties of the sludge decreases, the less opportunities for the MW energy to be absorbed and converted into heat); so, a corresponding increase in the penetration depth would be observed. Therefore, as the sludge gets dried the MW radiation progressively penetrates the sludge within deeper layers; thereby, contributing to achieve a uniform distribution of the MW power density as the drying process occurs.

Notably, increasing the thickness of the sludge induced a non-uniform heating behaviour inside the sludge sample. However, this effect was lessened by the reduction of the moisture content lowering the dielectric properties of the sludge and creating sort of a levelling effect regarding the achievement of a uniform heating profile. The latter can be seen as a competitive advantage exhibited by the MW dryers compared to the conventional dryers. That is, in traditional convective and conductive dryers the material thickness has a substantial effect on the drying performance. For instance, Gupta et al. (2012) when convective (hot air) drying cauliflower reported that the exposure time increased from 525 to 960 minutes when increasing the sample thickness from 50 to 160 mm to reach the target drying goal. Similar observations were reported by other authors (Ocoro-Zamora et al., 2013; Gojiya et al., 2015; Boutelba et al., 2018). That is, the system throughput capacity of conventional dryers in a similar range of evaluated material thickness as in this research increased by 82% compared to the only 14% observed in this research when MW drying. Thus, such levelling effect observed due to the moisture removal and their negative impact on the dielectric properties in MW drying brings important and differential advantages over conventional drying systems; such advantages may impact on higher throughput capacity and lower footprint for the MW drying systems compared to the conventional ones.

Furthermore, as seen in Figure 20b, the same three drying phases as previously observed in the previous Section 3.3.1.1 (i.e., a drying rate adaptation period, a constant rate drying period, and a falling rate drying period) were also noticed when modifying the thickness of the sludge. Similar observations as previously mentioned applied also to the trends observed in this section.

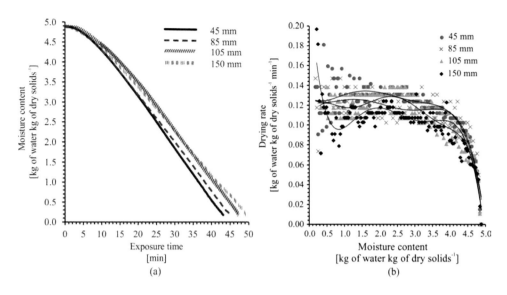

Figure 20: Effect of sludge sample thickness on: (a) the sludge moisture content as a function of the exposure time; and (b) the drying rates as a function of sludge moisture content

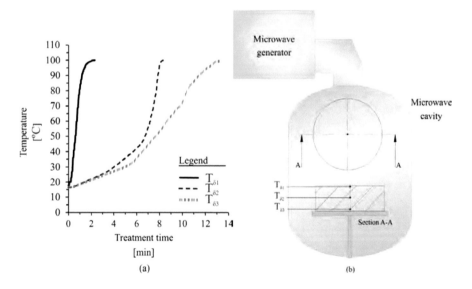

Figure 21: Effect of the sample thickness during the MW drying of the sludge on: (a) the temperature evolution profile at selected depths in the sludge sample and (b) a schematic representation of the temperature measurement points within the sludge sample

3.3.2 MW sludge drying performance: energy performance

This section discussed the effects of the sludge thickness and sludge mass on the energy performance of the MW drying system; main parameters under discussion in this section include the energy efficiency (μ_{en}) and the specific energy consumption (SEC). Both parameters provide an indication of the amount of energy that a given MW system would require to achieve a desired drying performance.

3.3.2.1 Energy efficiency

The energy efficiency was calculated for the evaluated sludge samples as described in Equation 11. The first set of experiments (experiments 1 to 4 as in Table 3) were conducted at the same sludge thickness and MW power output changing the initial mass of sludge from 1 to 4.5 kg. The second set of samples (experiments 5 to 8) were conducted at the same initial sludge mass and MW power output changing the sludge thickness from 45 to 150 mm. Figure 22 shows the effects of initial sludge mass (Figure 22a) and thickness (Figure 22b) on the energy efficiency as a function of moisture content. The trends of the energy efficiency curves for the evaluated samples relates well with the changes in drying rates previously reported. At the beginning of the drying process, an adaptation drying rate can be observed. Within this period, a little amount of water was evaporated; however, the system still receives a lot of energy in that period. Therefore, the energy efficiencies at the beginning of the drying period were very low (right side of Figure 22a and Figure 22b). As described in the Equation 11, the numerator includes both the energy required for the system for rising the temperature of the water (specific heat) and the energy needed for evaporating such water (latent heat). The denominator reflects the energy supplied in a certain exposure time which was related to the power supply. At the beginning of the drying process most of the energy was used raising the temperature of the sludge with none or little water evaporating at that time. Then, the numerator of Equation 11 is relatively low since the specific heat of the water is much lower than the latent heat. So lower efficiencies were observed and little amount of water was evaporated. When the adaptation phase period was finalized, the energy required to remove the water from the sludge was related to the latent heat of the water; therefore, the reported energy efficiencies were more or less constant in that period (constant rate drying period). The final phase (falling rate period) was not very noticeable in Figure 22. The changes in the energy efficiency during the final falling rate drying period were insignificant.

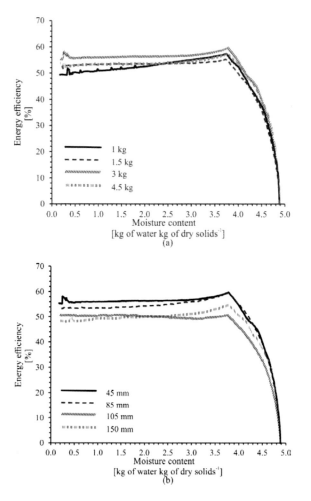

Figure 22: Effect of initial sludge mass (a) and thickness (b) on the energy efficiency as a function of the sludge moisture content

The results shown in Figure 22 also indicate that both the initial sludge mass (22a) and the thickness (22b) had a considerable impact on the overall energy efficiency of the system. For instance, as shown in Figure 22a, the energy efficiency increased with a decrease in applied load from 4.5 to 3 kg, after which it declined. Several researchers have reported that an increase in MW power density had a positive effect on both the system throughput capacity and on the energy performance of the system (Mawioo et al., 2016a; 2016b; Kocbek et al., 2020). That would be the trend that we were also expecting to see in this figure as well. However, a possible explanation for not seeing such a clear trend as reported by other authors could be due to the reflected power. An increase in the reflected energy by the irradiated material could reduce the amount of the power absorbed within the material; thus, with strong negative (or fluctuating) consequences on the energy efficiency (Atuonwu et al., 2019). Atuonwu et al. (2019) reported that the amount of MW power delivered to the system that is absorbed by the material increased with a decrease in the initial mass of the irradiated sample until a limit is reached; beyond that limit, the MW reflection increases, due to the unmatched condition. The reflected energy may thereafter be absorbed by any other component of the system such as the MW generator, the MW cavity, or it can be dissipated with the vapor/condensate (Kocbek et al., 2020). The energy performance of the MW system is directly related to the ability of the irradiated sample to absorb and convert the MW energy into heat. Thus, reducing the reflected power is fundamental for optimizing the performance of the system; and that would depend mostly on the operational parameters (such as MW power output, MW generation efficiency, shape of the sludge samples) and the amount of the initial sample mass irradiated (Ryynänen, 2002; Jafari et al., 2018; Atuonwu et al., 2019).

As observed in Figure 22b, the energy efficiency was much higher for the sludge samples having the lowest thickness of all the evaluated samples. For instance, at the end of the drying process energy efficiencies of 55, 53, 50 and 48% were reported for the sludge thickness of 45, 85, 105, and 150 mm, respectively. Such finding can be explained considering the non-uniformity of the temperature distribution within the material discussed in Section 3.3.1.2. The MW energy absorbed by the sample exponentially decreased with the thickness of the sample. In case the thickness of the material is larger than the MW penetration depth, larger non-uniform heating is expected; thus, both the system throughput capacity and the energy efficiency can be negatively affected. Notably, as discussed in previous Section 3.3.1.2 in MW drying, as a competitive advantage to conventional drying system, the MW penetration depth limitation can be counteracted to a certain extent by the reduction in the moisture content of the irradiated sample. Nonetheless, as suggested by Jafari et al. (2017), the thickness of the irradiated material should be as close as possible to the MW penetration depth.

3.3.2.2 Specific energy consumption

The SEC is defined as the energy consumed for removing one litre of water and is calculated as presented in the Equation 10. The SEC was calculated for the two set of experiments conducted (experiments 1 to 4 and 5 to 8 as described in Table 3) for the different initial mass of sludge at constant thickness and MW power output, and for the different thickness at constant initial sludge mass and MW power output. The results are presented in Figure 23. The overall SEC was calculated and reported when completely drying the sludge samples from an initial moisture content of 4.88 to a final moisture content of 0.18 kg of water kg of dry solids^{-1} (i.e., from 17% to 85% DS). The higher the energy efficiency, the lower the SEC. When discussing the set of experiments carried out changing the initial sludge mass, the lowest SEC was obtained when treating the sludge sample with an initial mass of 3 kg exhibiting a SEC of 4.93 MJ L^{-1} of water evaporated (1.37 kWh L^{-1}). Treating sludge with an initial sludge either higher or lower than 3 kg, resulted in higher SEC. As discussed in the energy efficiency sub-section, opposite trends were expected based on previous research carried out by the author Kocbek et al. (2020) and others (Sharma et al., 2006; Alibas, 2007; Aghilinategh et al., 2015). That is, it was expected to have the lowest SEC at the lowest irradiated initial sludge mass, since the power density would be the highest. This was already discussed earlier in this manuscript (Chapter 2). The different trends observed in this research could also be attributed to the reflective power mentioned in the energy efficiency discussion. The presence of energy that was not absorbed by the sludge, but rather dissipated somewhere else in the system would have a negative impact on the energy efficiency, also affecting the SEC. Probably the reflective energy for the 1 kg initial sludge sample was considerably higher than the reflective power for the other sludge initial mass samples; therefore, explaining the high SEC observed for this sample contradicting our previous findings.

Regarding the SEC as a function of the sludge thickness, an increase in the sludge thickness led to a reduction in the energy efficiency; therefore, an increase on the SEC as observed in Figure 23b. The non-uniform heating of the sludge sample as the thickness increased is the main reason for requiring higher SEC as the thickness layer increased. In conclusion, these findings highlight the critical role of operational parameters on absorption efficiency and uniformity of the MW energy within the material. By optimizing the sample mass load and thickness, both the efficiency of the MW absorption and the distribution of MW energy may be increased, leading to an enhanced energy performance of the system.

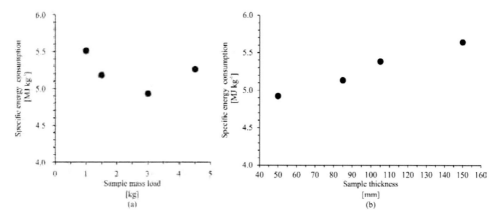

Figure 23: Effect of the initial sludge sample mass (a) and the thickness (b) on the overall specific energy consumption

3.4 Conclusions

- The system throughput capacity of the MW pilot scale system is governed by the rate at which the energy is absorbed per unit of sample volume (i.e., power density). Accordingly, the performance of the MW drying system unit can be enhanced by adjusting the power density across the sample. This can be achieved by changing the power to mass ratio. Operating at high power to mass ratios can lead to high drying rates (short exposure times). However, this may also give rise to energy losses in the form of reflective power, leading to a reduction in the energy efficiency of the system and high SEC.

- The non-uniform distribution of the electromagnetic energy within the material affects the energy and drying performance of the system. Such non-uniform behaviour is influenced by the thickness of the material. Increasing the sludge thickness layer exposed to radiation beyond the MW penetration depth has a negative impact on the performance of the system; the larger the sludge layer thickness, the greater the non-uniform distribution of the electromagnetic energy within the material leading to an increase in the exposure time, in the SEC, and a reduction in the energy efficiency. The MW penetration depth limitation was lessened by the removal of the sludge moisture decreasing the dielectric properties of the irradiated material. The latter effect offers some advantages to the MW drying compared to conventional drying systems.

- The results obtained in this research show that the efficiency and uniform distribution of the MW energy absorbed by the sludge over a given time interval is a key parameter in enhancing the energy performance of the system. In order to achieve a good heating uniform distribution with minimum loses, an optimal sample thickness, mass, and MW parameters needs to be provided.

References

Aghilinategh N, Rafiee S, Hosseinpour S, Omid M, Mohtasebi SSJFS, Nutrition. (2015). Optimization of intermittent microwave–convective drying using response surface methodology. 3(4), 331-341.

Alibas I. (2007). Energy Consumption and Colour Characteristics of Nettle Leaves during Microwave, Vacuum and Convective Drying. Biosystems Engineering, 96(4), 495-502.

Antunes E, Jacob MV, Brodie G, Schneider PA. (2018). Microwave pyrolysis of sewage biosolids: Dielectric properties, microwave susceptor role and its impact on biochar properties. Journal of Analytical and Applied Pyrolysis, 129, 93-100.

Antunes E, Schumann J, Brodie G, Jacob MV, Schneider PA. (2017). Biochar produced from biosolids using a single-mode microwave: Characterisation and its potential for phosphorus removal. Journal of environmental management, 196, 119-126.

Atuonwu J, Tassou S. (2019). Energy issues in microwave food processing: A review of developments and the enabling potentials of solid-state power delivery. Critical reviews in food science and nutrition, 59(9), 1392-1407.

Azimi-Nejadian H, Hoseini SS. (2019). Study the effect of microwave power and slices thickness on drying characteristics of potato. Heat and Mass Transfer, 55(10), 2921-2930.

Bennamoun L, Arlabosse P, Léonard A. (2013). Review on fundamental aspect of application of drying process to wastewater sludge. Renewable and Sustainable Energy Reviews, 28, 29-43.

Bennamoun L, Chen Z, Afzal MT. (2016). Microwave drying of wastewater sludge: Experimental and modeling study. Drying Technology, 34(2), 235-243.

Bermúdez J, Beneroso D, Rey-Raap N, Arenillas A, Menéndez J. (2015). Energy consumption estimation in the scaling-up of microwave heating processes. Chemical Engineering and Processing: Process Intensification, 95, 1-8.

Bhattacharya M, Basak T. (2016). A review on the susceptor assisted microwave processing of materials. Energy, 97, 306-338.

Boutelba I, Zid S, Glouannec P, Magueresse A, Youcef-ali S. (2018). Experimental data on convective drying of potato samples with different thickness. Data in Brief, 18, 1567-1575.

Budd CJ, Hill AJIjoh, transfer m. (2011). A comparison of models and methods for simulating the microwave heating of moist foodstuffs. 54(4), 807-817.

Çelen S. (2019). Effect of microwave drying on the drying characteristics, color, microstructure, and thermal properties of trabzon persimmon. Foods, 8(2), 84.

Chandrasekaran S, Ramanathan S, Basak TJAJ. (2012). Microwave material processing—a review. 58(2), 330-363.

Chen Z, Afzal MT, Salema AA. (2014). Microwave drying of wastewater sewage sludge. Journal of Clean Energy Technologies, 2(3), 282-286.

Christodoulou A, Stamatelatou K. (2016). Overview of legislation on sewage sludge management in developed countries worldwide. Water Science and Technology, 73(3), 453-462.

Comission E. (2008). Directive 2010/75/EU of the European Parliament and of the Council of 24 November 2010 on industrial emissions (integrated pollution prevention and control) Text with EEA relevance.

Darvishi H, Mohamamdi P, Azadbakht M, Farhudi Z. (2018). Effect of different drying conditions on the mass transfer characteristics of kiwi slices. Journal of Agricultural Science and Technology, 20(2), 249-264.

Dominguez A, Menéndez J, Inguanzo MP. (2004). Sewage sludge drying using microwave energy and characterization by IRTF. Afinidad, 61(512), 280-285.

Flaga A. (2005). *Sludge drying.* Paper presented at the Proceedings of Polish-Swedish seminars, Integration and optimization of urban sanitation systems. Cracow March.

Fytili D, Zabaniotou A. (2008). Utilization of sewage sludge in EU application of old and new methods—a review. Renewable and sustainable energy reviews, 12(1), 116-140.

Gaukel V, Siebert T, Erle U. (2017). 8 - Microwave-assisted drying. In Regier M, Knoerzer K, Schubert H (Eds.), The Microwave Processing of Foods (Second Edition) (pp. 152-178): Woodhead Publishing.

Gojiya D, Vyas DJJoFP, Technology. (2015). Studies on effect of slice thickness and temperature on drying kinetics of Kothimbda (Cucumis callosus) and its storage. 6(1).

Guilong W, Guoqun Z, Huiping L, Yanjin G. (2010). Analysis of thermal cycling efficiency and optimal design of heating/cooling systems for rapid heat cycle injection molding process. Materials & Design, 31(7), 3426-3441.

Gupta M, Leong EWW. (2007). Microwaves and metals: John Wiley & Sons.

Gupta MK, Sehgal V, Kadam DM, Singh A, Yadav YJAJoB. (2012). Effect of bed thickness on cauliflower drying. 2(5), 56-61.

Haque KE. (1999). Microwave energy for mineral treatment processes—a brief review. International Journal of Mineral Processing, 57(1), 1-24.

Herzel H, Krüger O, Hermann L, Adam C. (2016). Sewage sludge ash—A promising secondary phosphorus source for fertilizer production. Science of the Total Environment, 542, 1136-1143.

Jafari H, Kalantari D, Azadbakht M. (2017). Semi-industrial continuous band microwave dryer for energy and exergy analyses, mathematical modeling of paddy drying and it's qualitative study. Energy, 138, 1016-1029.

Jafari H, Kalantari D, Azadbakht M. (2018). Energy consumption and qualitative evaluation of a continuous band microwave dryer for rice paddy drying. Energy, 142, 647-654.

Jakobsson C. (2014). Sustainable agriculture: Baltic University Press.

Keey R. (1991). Drying of loose and particulate materials: CRC Press.

Kelessidis A, Stasinakis AS. (2012). Comparative study of the methods used for treatment and final disposal of sewage sludge in European countries. Waste management, 32(6), 1186-1195.

Khan MImran H, Welsh Z, Gu Y, Karim MA, Bhandari B. (2020). Modelling of simultaneous heat and mass transfer considering the spatial distribution of air velocity during intermittent microwave convective drying. International Journal of Heat and Mass Transfer, 153, 119668.

Kocbek E, Garcia HA, Hooijmans CM, Mijatović I, Lah B, Brdjanovic D. (2020). Microwave treatment of municipal sewage sludge: Evaluation of the drying performance and energy demand of a pilot-scale microwave drying system. Science of The Total Environment, 742, 140541.

Köhler C, Venditti S, Igos E, Klepiszewski K, Benetto E, Cornelissen A. (2012). Elimination of pharmaceutical residues in biologically pre-treated hospital wastewater using advanced UV irradiation technology: a comparative assessment. Journal of hazardous materials, 239, 70-77.

Kumar C, Joardder M, Farrell TW, Karim M. (2016a). Multiphase porous media model for intermittent microwave convective drying (IMCD) of food. International Journal of Thermal Sciences, 104, 304-314.

Kumar C, Joardder M, Farrell TW, Karim M. (2016b). Multiphase porous media model for intermittent microwave convective drying (IMCD) of food. International Journal of Thermal Sciences, 104, 304-314.

LeBlanc RJ, Matthews P, Richard RP. (2009). Global atlas of excreta, wastewater sludge, and biosolids management: moving forward the sustainable and welcome uses of a global resource: Un-habitat.

Ledakowicz S, Stolarek P, Malinowski A, Lepez O. (2019). Thermochemical treatment of sewage sludge by integration of drying and pyrolysis/autogasification. Renewable and Sustainable Energy Reviews, 104, 319-327.

Liu CM, Wang QZ, Sakai N. (2005a). Power and temperature distribution during microwave thawing, simulated by using Maxwell's equations and Lambert's law. International journal of food science & technology, 40(1), 9-21.

Liu CM, Wang QZ, Sakai NJIjofs, technology. (2005b). Power and temperature distribution during microwave thawing, simulated by using Maxwell's equations and Lambert's law. 40(1), 9-21.

Mawioo PM, Garcia HA, Hooijmans CM, Velkushanova K, Simonič M, Mijatović I, Brdjanovic D. (2017). A pilot-scale microwave technology for sludge sanitization and drying. Science of the Total Environment, 601, 1437-1448.

Mawioo PM, Hooijmans CM, Garcia HA, Brdjanovic D. (2016a). Microwave treatment of faecal sludge from intensively used toilets in the slums of Nairobi, Kenya. Journal of Environmental Management, 184, 575-584.

Mawioo PM, Rweyemamu A, Garcia HA, Hooijmans CM, Brdjanovic D. (2016b). Evaluation of a microwave based reactor for the treatment of blackwater sludge. Science of the Total Environment, 548, 72-81.

Mello PA, Barin JS, Guarnieri RA. (2014). Chapter 2 - Microwave Heating. In Flores ÉMdM (Ed.), Microwave-Assisted Sample Preparation for Trace Element Analysis (pp. 59-75). Amsterdam: Elsevier.

Menendez J, Inguanzo M, Pis J. (2002). Microwave-induced pyrolysis of sewage sludge. Water research, 36(13), 3261-3264.

Mujumdar AS, Devahastin S. (2000). Fundamental principles of drying. Exergex, Brossard, Canada, 1(1), 1-22.

Ni H, Datta A, Torrance K. (1999). Moisture transport in intensive microwave heating of biomaterials: a multiphase porous media model. International Journal of Heat and Mass Transfer, 42(8), 1501-1512.

Ocoro-Zamora MU, Ayala-Aponte AAJD. (2013). Influence of thickness on the drying of papaya puree (Carica papaya L.) through Refractance WindowTM technology. 80(182), 147-154.

Pitchai K, Birla SL, Subbiah J, Jones D, Thippareddi H. (2012). Coupled electromagnetic and heat transfer model for microwave heating in domestic ovens. Journal of Food Engineering, 112(1), 100-111.

Ryynänen S. (2002). Microwave heating uniformity of multicomponent prepared foods.

Schaum C, Lux J. (2010). Sewage sludge dewatering and drying. ReSource–Abfall, Rohstoff, Energie, 1, 727-737.

Sharma GP, Prasad S. (2006). Specific energy consumption in microwave drying of garlic cloves. Energy, 31(12), 1921-1926.

Soltysiak M, Erle U, Celuch M. (2008). *Load curve estimation for microwave ovens: experiments and electromagnetic modelling.* Paper presented at the MIKON 2008-17th International Conference on Microwaves, Radar and Wireless Communications.

Sólyom K, Mato RB, Pérez-Elvira SI, Cocero MJ. (2011). The influence of the energy absorbed from microwave pretreatment on biogas production from secondary wastewater sludge. Bioresource Technology, 102(23), 10849-10854.

Stuchly S, Hamid M. (1972). Physical parameters in microwave heating processes. Journal of Microwave Power, 7(2), 117-137.

Stuerga D. (2006). Microwave-material interactions and dielectric properties, key ingredients for mastery of chemical microwave processes (Vol. 2): WILEY-VCH Verlag GmbH & Co. KGaA.

Thostenson E, Chou T-W. (1999). Microwave processing: fundamentals and applications. Composites Part A: Applied Science and Manufacturing, 30(9), 1055-1071.

Zhu H, Gulati T, Datta AK, Huang K. (2015). Microwave drying of spheres: Coupled electromagnetics-multiphase transport modeling with experimentation. Part I: Model development and experimental methodology. Food and Bioproducts Processing, 96, 314-325.

4

Microwave treatment of municipal sludge

Effects of the sludge physical-chemical properties on the microwave drying performance

This chapter is based on: Kocbek E, Garcia HA, Hooijmans CM, Mijatović I, Kržišnik D, Humar M, Brdjanovic D. Effects of the sludge physical-chemical properties on the microwave drying performance of the sludge. Submitted to Journal of Science of The Total Environment, 2021.

Abstract

Thermal drying is an effective sludge treatment method for dealing with large volumes of sludge. Microwave (MW) technology has been proposed as an effective and efficient technology for sludge drying. The physical-chemical properties of the sludge depend both on the origin of the sludge, as well as on the treatment process at which the sludge has been exposed. The physical-chemical properties of the sludge affect the performance and the subsequent valorisation and management of the sludge. This study evaluated the effect of certain physical-chemical properties of the sludge (moisture content, organic content, calorific value, porosity, hydrophobicity, and water-sludge molecular interaction, among others) on the MW sludge drying and energy performance. Four different types of sludge were evaluated collected from municipal wastewater treatment plants and for septic tanks. The performance of the MW system was assessed by evaluating the sludge drying rates, exposure times, energy efficiencies and power input consumed by the MW system and linking the MW drying performance to the sludge physical-chemical properties. The results confirmed that MW drying substantially extends the constant drying period associated with unbound water evaporation, irrespective of the sludge sample evaluated. However, the duration and intensity were determined to depend on the dielectric properties of the sludge, particularly on the distribution of bound and free water. Sludge samples with a higher amount of free and loosely bound water absorbed and converted MW energy into heat more efficiently than sludge samples with a lower amount of free water. As a result, the sludge drying rates increased and the constant drying rate period prolonged; hence, leading to an increase in MW drying energy efficiency. The availability of free and loosely bound water molecules was favoured when hydrophobic compounds, e.g., oils and fats, were present in the sludge.

4.1 Introduction

On-site sanitation systems and municipal wastewater treatment plants (WWTPs) generate significant amounts of sludge which needs to be treated before its final disposal or reuse (Kelessidis et al., 2012). The sludge, if not correctly managed, can eventually harm the natural resources and affect public health. However, if properly treated several valuable resources can be recovered (Kehrein et al., 2020). The sludge contains large amounts of organic matter, nitrogen, and phosphorus. Nonetheless, the sludge may also contain pathogens and contaminants such as heavy metals and organic substances which may limit its reuse as a fertilizer. Particularly, municipal sludge receiving contributions from industries and hospitals may exhibit some of these issues (Kroiss et al., 2007). Therefore, both the sludge characteristics, as well as the sludge management strategies determine the subsequent reuse or disposal possibilities of the sludge and the potential environmental and/or public health impacts (Đurđević et al., 2020). Typically strategies commonly used to manage/treat the sludge include (co-) incineration, (co-) composting, and agricultural amendment, among others (Gomes et al., 2019). In all cases, there are both significant energy expenditures, as well as management issues due to the elevated water content of the sewage sludge with sludge moisture contents as high as approximately 98% (2% dry solids (DS)). Typically, sludge moisture contents for treated sewage sludge of 60% (40% DS) and 15% (85% DS) should be reached for agricultural reuse and for energy recovery applications such as (co-)incinerations, respectively (Kamran et al., 2020; Lam et al., 2020). The high moisture content of the raw sludge also influences its collection (due to the large sludge volumes requiring collection), the transportation costs, and the frequency at which the sludge needs to be collected, among others. Thus, the high moisture content of the raw sludge impacts on both the provision of proper sludge management strategies, as well as on the related treatment costs. In recent years, mechanical and thermal dewatering methods have been proposed as suitable alternatives for sludge treatment (Flaga, 2005). Mechanical dewatering can reduce the sludge moisture content up to approximately 70% (30% DS) and thermal drying up to approximately 5% (95% DS). However, currently available mechanical and thermal dewatering/drying systems exhibit several disadvantages including: high demands of energy, inefficient treatment processes, and large exposure times required for achieving the treatment goals (Ohm et al., 2009; Mujumdar, 2014).

Microwave (MW) irradiation has been presented as a viable technology for sludge drying achieving sludge volume reductions higher than 90% at much shorter exposure times compared to traditional drying methods such as hot air-drying techniques (Dominguez et al., 2004; Kocbek et al., 2020). MWs are electromagnetic waves in the range between 37.24 and 12.24 cm (i.e., frequencies of 915 and 2450 MHz, respectively) (Haque, 1999; Bilecka et al., 2010). The primary mechanism by which nonionizing electromagnetic energy is converted into heat during MW treatment is attributed to the dipolar polarisation of certain compounds in the irradiated material (Stuerga, 2006). The electromagnetic field causes rotational motion of molecules with a permanent dipole moment, such as the water molecules in the sludge (Stuerga, 2006). The water molecules attempt to resist the oscillating field's changes, resulting in friction and heat, which occurs throughout the material (Mishra et al., 2016). In such a way, the electromagnetic energy is converted into heat. The heat transfer process is regulated by the dielectric loss tangent (tan δ) of the irradiated material defined as the ratio between the dielectric loss factor (ε") and the dielectric constant (ε'). The dielectric loss factor represents the conversion of the electromagnetic energy into heat, while the dielectric constant depicts the ability of the material to store electromagnetic energy. Therefore, a material characterized by a high dielectric loss factor and a low dielectric constant (i.e., a high tan δ), such as the sludge, can be efficiently heated by applying MW radiation (Antunes et al., 2018). The tan δ of the irradiated material is a relevant factor when applying MW energy, and this property may change among the various MW irradiated materials. Particularly, the tan δ may significantly change depending on the physical-chemical characteristics of the sludge such as the sludge moisture and organic content.

In fact, Mawioo et al. (2017) irradiated municipal sludge, septic tank sludge (SS), and fresh faecal sludge aiming at sludge sterilization and drying. The authors reported complete bacterial inactivation and sludge drying up to a final moisture content of 5% (95% DS). Moreover, the authors reported a strong dependence of the MW performance on the physical-chemical characteristics of the sludge. For instance, the fresh faecal sludge samples, exhibiting the lowest initial moisture content of all the evaluated sludge samples of 77% (23% DS), were dried much faster and demanded less energy compared to the other evaluated types of sludge. Therefore, the initial moisture content seemed to be one of the key physical-chemical sludge properties when MW drying the sludge. However, the authors also reported faster temperature increments and faster water evaporation rates when irradiating waste activated sludge (WAS) compared to when irradiating SS. The moisture content of the WAS was much higher than the moisture content of the SS; then, showing opposite trends as previously described. The authors suggested that the disparities concerning the drying rates and exposure times could be attributed not only to the moisture content, but also to the organic matter content, which was higher in WAS (75%) than in SS (55%). Mawioo et al. (2017) hypothesized that the organic compounds present in the sludge such as carbohydrates and proteins could exhibit high dielectric loss tangent factors, leading to an increase in the amount of MW energy converted into heat. However, the organic matter (mostly carbohydrates and proteins) have also been reported to exhibit a negative impact on the MW drying performance (Dealler et al., 1992; Léonard et al., 2004). The organic matrix present in the sludge is composed of organic compounds which some of them may exhibit hydrophilic and polar groups increasing the chances of water molecules to strongly bind to such groups. Thus, such a strong bonding between the water molecules and the organic components in the sludge counteract the effects caused by the MW radiation related to promoting the rotation of the free (not bound) water molecules (Jones et al., 2003; Pickles et al., 2014). A fraction of the water molecules present in the organic matrix in the sludge would exist in a relatively free liquid form; however, some of the water molecules would be attached to the organic matrix of the sludge tightly bound to the sludge (Vesilind, 1994). Thus, such strong binding would hinder the rotation of the water molecules, negatively impacting the conversion of the electromagnetic energy into heat; so, making the MW drying process less efficient (Jones et al., 2003; Pickles et al., 2014). As such knowing precisely the physical-chemical properties of the sludge such as the water content and the organic content (among others) would contribute to better predict the MW drying performance of different types of sludge.

In addition, the affinity of the sludge for water and the strength at which the water molecules are bound to the organic matrix of the sludge can be better assessed by determining the water sorption/desorption properties of the sludge (i.e., the moisture sorption isotherms). The water sorption/desorption properties of the sludge could have a strong impact both on the sludge dehydration processes, as well as on the stability of the dried sludge for storage purposes. Depending on the relative humidity (RH) (i.e., the ratio between the water vapour pressure in the air and the saturation water vapour pressure in the air) at which a material is exposed, the material can adsorb or desorb water from the atmosphere gaining or losing moisture (Aviara, 2020). When the material no longer adsorbs or releases water from or to the environment, the material is at an equilibrium; that equilibrium stage is known as the equilibrium moisture content (EMC). The material either gains or loses moisture from its surroundings to reach the EMC depending on whether the ambient vapour pressure is higher or lower than the equilibrium vapour pressure (Aviara, 2020). The graphical representation of the water content of the material as a function of the RH at EMC conditions and at constant temperature is referred to as the moisture sorption isotherm. The moisture sorption isotherm provides insight into the sludge moisture-binding characteristic, which may vary according to: (i) the physical-chemical composition of the sludge, (ii) the structure of the sludge, and (iii) the process used to dry the sludge. Furthermore, from the sorption isotherms, the isosteric heat of sorption (a thermodynamic parameter) can be determined. The isosteric heat of sorption offers insight into the energy of the forces involved in the chemical bonding between the sludge and the water molecules (Poyet et al., 2009; Souza et al., 2015). That is, the energy required to break the intermolecular attractive forces between the water vapour and the sludge (i.e., the isosteric heat of sorption) can be determined from the sorption/desorption properties of the sludge (moisture sorption isotherms). Knowing the isosteric heat of sorption is highly desirable for better understanding the actual energy needs for sludge drying, avoiding over-drying of the sludge while minimizing the total energy consumption (i.e., the treatment costs).

The physical-chemical characteristics of the sludge are determined by both the origin of the sludge, and by the treatment processes at which the sludge has gone through. Such characteristics of the sludge strongly influence subsequent sludge drying processes such as the MW sludge drying. Particularly, the physical-chemical properties will determine the efficacy of the MW treatment for the sanitization and drying of the different types of sludge, as well as the energy expenditures involved in such processes. The effects of the sludge properties on the MW drying performance have not been thoroughly reported in the literature; therefore, they need to be assessed. This research addressed such needs directly. This research aims firstly at determining the physical-chemical properties of different types of sludge originated from different municipal WWTPs and on-site facilities; such sludge properties were also related to the origin of the sludge and to the wastewater treatment processes at which the sludge was exposed. The sludge physical-chemical properties determined in this study included: moisture content, organic content, elemental composition, calorific value, absolute and envelope density, porosity, oil and grease content, presence of heavy metals, water sorption/desorption properties, and isosteric heat of sorption. Secondly, this study aims at determining the impact of such physical-chemical properties of the sludge on the MW drying performance. The MW drying performance was assessed by determining: overall sludge drying performance (exposure times to achieve a certain degree of drying), drying rates, specific energy inputs, and energy efficiencies.

4.2 Materials and methods

The physical-chemical properties of different types of sludge (municipal sewer sludge and septic tank sludge generated) were characterized including the determination of moisture content, organic content, elemental composition, calorific values, absolute and bulk density, porosity, oil and grease content, heavy metals, water sorption isotherms, and isosteric heat of sorption. Subsequently, the drying performance of the different types of sludge was evaluated in a pilot-scale MW system operated at MW output power of 6 kW. The MW drying performance was assessed by evaluating the exposure times required to reduce the moisture content of the sludge samples to approximately 85% (15% DS), the energy efficiencies, and the specific energy inputs. Finally, the effect of the physical-chemical properties of sludge were related to the MW drying performance.

4.2.1 Materials and sample preparation

Three different types of WAS were obtained from three different WWTPs, while one type of SS was obtained from an onsite sanitation facility. Table 4 describes the origin of the sludge as well as the type of treatment technologies producing each particular type of sludge.

The WWTPs A and B (as described in Table 4) were equipped with sequential batch reactors for the treatment of the wastewater collected from both domestic and industrial sources. The WWTP C was equipped with a conventional activated sludge process for the treatment of similar wastewater as in WWTPs A and B. The WAS samples were mechanically dewatered by centrifuges; polymers were added to aid the dewatering process. The dewatered WAS samples were collected for conducting the experiments described in this study.

The SS was obtained from septic tanks from households located in remote rural areas without access to sewers. The septic tanks were only receiving domestic wastewater discharges. The SS was transported in septic sludge trucks to a local WWTP in Ptuj (Slovenia), where it was sieved into an equalization tank (Figure 24), prior its co-treatment with municipal wastewater. The SS samples were collected from a SS equalisation tank. The SS samples were further destabilized and flocculated by the addition of 40 mg L^{-1} of a cationic based polymer (Acefloc 80902+, Allied Solutions). The resulting SS samples were further sieved through a 0.5 mm mesh size sieve. The study was conducted during the winter months. During the winter period, this plant received SS collected from household locations due to emergencies such as clogging and SS tank overspill. Such sludge could be retained in the septic tanks at household locations up to approximately five years or longer. Therefore, the SS samples evaluated in this research could have been retained in the septic tanks for a long time before sampling. All the sludge samples were stored in plastic containers at 4 °C and evaluated and/or analysed within 48 h.

Figure 24: Septic tank sludge intake station at WWTP (Ptuj, Slovenia)

Table 4: WWTPs and onsite sanitation facilities where the sludge samples were collected

Sludge samples		WWTP location	Treatment technology	Wastewater/Sludge source	Mechanical pre-treatment process	Biological treatment process	Sludge treatment process
WAS	A	Ptuj	Sequential batch reactor	Industrial and domestic (80%)	Screening, grit and grease removal	Anaerobic, anoxic and aerobic	Mechanical dewatering using centrifuge assisted by the addition of polymers
	B	Ljutomer	Sequential batch reactor	Industrial and domestic (80%)	Screening, grit and grease removal	Anaerobic, anoxic and aerobic	Mechanical dewatering using centrifuge assisted by the addition of polymers
	C	Maribor	Conventional waste activated sludge	Industrial and domestic (80%)	Screening, grit and grease removal	Anaerobic, anoxic and aerobic	Saturated air flotation and mechanical dewatering using centrifuge assisted by the addition of polymers
SS	D	Ptuj	Septic tank	Domestic	Screening	Anaerobic	Gravitational draining assisted by the addition of polymers

4.2.2 Analytical procedures

4.2.2.1 Total solids (TS) and volatile solid (VS) determination

The TS and VS were determined according to the gravimetric methods SM-2540D and SM-2540E, as described in the American Public Health et al. (2005). The DS% is the same as the TS concentration expressed in percentage; the moisture content was calculated subtracting the DS% to 100.

4.2.2.2 Calorific value and Carbon (C), Hydrogen (H), Nitrogen (N), and Sulphur (S) determinations

Carbon (C), hydrogen (H), nitrogen (N), and sulphur (S) are the main chemical elements that determine the energy content of sludge (i.e., the gross calorific value of the sludge) (Friedl et al., 2005). The gross calorific values, and the C, H, N and S content of the sludge samples were determined at the Institute of Chemistry, Ecology, Measurement and Analytics (IKEMA, Lovrenc na Dravskem polju, Slovenia) according to the following methods: (i) calorific value (SIST-TS CEN/TS 16023:2014 standard); (ii) elemental sulphur and hydrogen (Dumas method); (iii) total carbon (SIST EN 13137:2002); and (iv) total nitrogen (SIST EN 16168:2013). The gross calorific value of the sludge was determined by measuring the amount of heat emitted during the complete combustion of the sludge in a bomb calorimeter (IKA-Calorimeter C 400 adiabatisch IKA®-Werke GmbH & Co. KG, Staufen German). The determined gross calorific value included the condensation enthalpy of the water. The net calorific value was obtained by subtracting the condensation enthalpy from the gross calorific value.

4.2.2.3 N-Hexane Extractable Material (HEM) determination

The presence of N-Hexane-extractable materials (HEM), often termed as oil and grease, as shown in Figure 25, were determined according to the gravimetric separatory funnel extraction method suggested on the EPA Method 1664, revision B. The sludge samples for the oil and grease determinations were prepared by diluting a known amount of a mechanically dewatered sludge sample in one litre of distilled water. A dispersing instrument was used to homogenize the diluted sample at a circumferential speed of 10,000 rpm for 10 minutes (Ultra-Turrax T 25, IKA, Slovenia). The pH of the sludge samples was adjusted to a value of below two by adding four mL of a concentrated sulfuric acid (H_2SO_4) solution. Then, the oil and grease extraction were performed by adding 50 mL of dichloromethane (solvent) continuously in a separatory funnel. The funnel was placed in a flat-bed rotary shaker for ten minutes at 400 rpm. The solvent was separated in test tubes and centrifuged at 3,000 rpm for five minutes. The solvent layer (supernatant) was passed through anhydrous sodium sulphate (Na_2SO_4) for drying. This process was repeated at least four times. The solvent was evaporated under a reduced pressure at 40 °C using a rotary evaporator. The residues of the oil and grease were weighed. The analysis was repeated twice, and the average of these two determinations was reported.

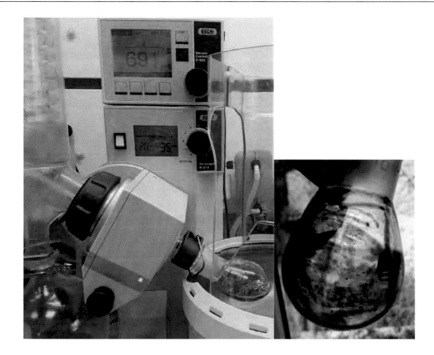

Figure 25: Rotary evaporator (left) and fats and oils extracted from WAS sample C (right)

4.2.2.4 X-ray Fluorescence

The X-ray fluorescence analyses were carried out to determine elements; specifically, heavy metals such as nickel (Ni), zinc (Zn), lead (Pb), chromium (Cr), and copper (Cu). The determinations were initiated by pelletizing the MW dried WAS and SS samples using a Chemplex Spectro pellet press (Chemplex Industries Inc., Palm City, USA) at a pressure of 12 tons (120 kilonewtons) in a five minutes period. The press generated sludge pellets with a radius of 16 mm and thickness of 10 mm. The concentrations of the elements (Ni, Zn, Pb, Cr, and Cu) in the pellets were determined using a Twin-X X-ray fluorescence spectrometer (XRF Twin-X, Oxford Instruments, Abingdon, UK). The system was equipped with photodiodes with a p–i–n semiconductor structure, and it was operated for 360 seconds at a voltage of 26 kV and at a current of 112 μA. The spectrums were analysed by the aid of the SmartCheck software provided in the instrument.

4.2.2.5 Determination of the absolute density, envelope (bulk) density, and sludge porosity

The MW dried samples were further dried in an oven at 40 °C for 72 hours to reach a constant mass. The absolute density of the sludge was determined in a helium atmosphere using an AccuPyc 1330 automatic gas pycnometer (Micromeritics Inc., USA). Before carrying out the volume measurements, the sludge samples were weighed. The sludge samples were introduced in the pycnometer chamber, and the chamber was purged ten times with helium gas to remove gases and moisture that could be present in the samples. The purged sample was subjected to a static pressure at an increasing pressure of 0.020 psig min^{-1} allowing the gas to enter the pores of the sludge. The instrument determined the gas displaced by the sample by applying the ideal gas law and in that way the absolute volumes of the samples (and the densities) were calculated (Candanedo et al., 2005).

The envelope density was determined by using a GeoPyc 1360 (Micromeritics Inc., USA) instrument. The volume of the samples was analysed by packing the sample in DryFlo silica sand (registered trademark, Micromeritics Inc., USA). Then, the sample chamber of the instrument was filled with only sand and the volume of the sand was calculated. Thereafter, the sludge sample packed with the sand was introduced into the chamber, and the total volume of the sand and sludge sample was measured. From the difference in volume, the sludge sample envelope volume and density were calculated.

The volume of the sludge occupied by the air is directly related to the porosity of sludge which denotes the volume of interstices in relation to the total volume of the material. The porosity of the sludge was determined as the ratio of pore space volume or voids within the sample (the difference between envelope and absolute density) to the envelope volume.

4.2.2.6 Determination of the water sorption/desorption properties of the sludge

The water sorption and desorption properties of the sludge were determined to obtain an insight into both the moisture-binding characteristic of the sludge solids, as well as on the distribution of the bound water within the sludge. The determination of the water sorption and desorption properties were carried out by finding the moisture sorption isotherms (i.e., the equilibrium relationship between the moisture content of the sample at equilibrium - at EMC - and the RH). Such moisture sorption isotherms were determined using a dynamic vapour sorption (DVS) intrinsic apparatus (DVS Intrinsic, Surface Measurement Systems Ltd., London, UK). The different MW dried sludge samples were initially conditioned for approximately 24 h at 20 ± 0.2 °C and at a RH of 1 ± 1% in a separate chamber exposed to a constant flow of dry air. Approximately, 40 mg of the conditioned sludge samples were placed in a holder and laid in the DVS apparatus provided with a microbalance within a sealed thermostatically controlled DVS chamber. A constant flow of dry air was passed over the sludge samples at a flowrate of 200 cm^3 s^{-1} and at a temperature of 25 ± 0.1°C at a RH range from 0 to 95%. The RH range was gradually increased/decreased at 5% RH steps increments between 0% and 95% RH for both the sorption and the desorption determinations. Two complete isotherms runs were performed for each sludge sample to capture both the sorption and desorption behaviour of the material (i.e., four runs per sample). The instrument maintained a constant target RH until the change in the sample moisture content as a function of time (dm/dt) was less than 0.002% per minute over a 10 min period. The running time, target RH, actual RH, vapour pressure, and sample weight were recorded by the instrument every 20 s throughout the isotherm run. The moisture sorption isotherms were obtained for every sludge sample by plotting the moisture content of the sample (once the equilibrium was reached at each evaluated RH) as a function of the RH. The results presented in this study considered the results obtained when carrying out the second sorption/desorption run.

4.2.2.7 Determination of isosteric heat of sorption

The total isosteric heat of sorption provides valuable information regarding the strength of the interaction between the water molecules and the material; thus, the energy that would be needed to completely dry the material. The total isosteric heat of sorption is defined as the sum of the heat change accompanying the isothermal sorption of a specified quantity of water vapour on the sludge sample (i.e., net heat of sorption), and the energy required for a normal water vaporisation (latent heat of evaporation of the unbound water). The net isosteric heat of sorption was determined as described in the Clausius-Clapeyron equation (Equation 19) (Poyet et al., 2009). The inputs for that equation were obtained using the same DVS instrument as previously described in Section 4.2.2.6. The net heat of sorption was calculated for each RH after the sludge sample reached moisture equilibrium. The very same procedure as described in Section 4.2.2.6 was followed with the sludge samples. The isotherms were carried out at two different temperatures (i.e., 25 °C and 40 °C).

$$h_{st} = Rln\left[\frac{p_1}{p_2}\right]\left(\frac{T_1T_2}{T_1 - T_2}\right) \qquad (19)$$

where h_{st} is the net heat of sorption [kJ kg^{-1}], R is the universal gas constant [kJ kg^{-1} °C^{-1}], and p_1 and p_2 [kPa] are the vapour partial pressure measured using the DVS instrument in a thermostatically sealed chamber at the two evaluated temperatures, T_1 and T_2 [°C].

The net heat of sorption is used for calculating the total isosteric heat of sorption determined according to the Equation 20 (Sousa et al., 2016):

$$H_{st} = h_{st} + H_v \qquad (20)$$

where H_{st} is the total isosteric heat of sorption [kJ kg^{-1}], and H_v is the latent heat of evaporation of pure water [kJ kg^{-1}]. The latent heat of evaporation of pure water was calculated as described in the Equation 21 (Sousa et al., 2016):

$$H_v = 2502.2 - 2.39T \qquad (21)$$

where T denotes the average temperature of the studied range (i.e., 25 °C and 40 °C). The total isosteric heat of sorption H_{st} was calculated for every RH when the sludge samples reached equilibrium regarding the moisture content. Due to the availability of the DVS instruments, only the total isosteric heat of sorption H_{st} for the WAS sample from the WWTP A was determined.

4.2.3 Experimental pilot unit

An experimental MW pilot-scale sludge drying system was designed and built for this research (Tehnobiro d.o.o, Maribor, Slovenia). A detailed schematic of the experimental pilot-scale MW system is shown in Figure 26. The MW system consisted of a MW power supply and a MW generator delivering a maximum MW power output of 6 kW operating at a frequency of 2,450 MHz. The MWs were directed to a stainless-steel drying cavity provided with a polypropylene (PP) turntable able to rotate at a speed of one rpm. The turntable was added to provide a uniform radiation distribution all over the irradiated material that would result in a better temperature distribution through all over the sludge sample. The sludge samples were placed in a cylindrical PP holding vessel with a maximum sludge capacity of 6 kg. The MW generator, connected to an isolator, delivered the electromagnetic energy to the drying cavity along a standard rectangular waveguide WR340 (86.36 × 53.18 mm). The isolator allowed the MW energy to be transmitted to the cavity (forward power) but not in the opposite direction (reflected power). A dummy water load conditioned with demineralized water at a flowrate of approximately 600 L h^{-1} was connected to the waveguide circulator used to absorb the reflected power and to prevent overheating of the MW generator. The demineralized water was also used for cooling down the MW power supply. The vapour generated during the MW drying process from the drying cavity was extracted to an odour filtration system. To selectively increase the adsorption capacity of the air filtration system, three different types of fillers were used. These fillers included activated carbon soaked in phosphoric acid (H_3PO_4), sodium hydroxide (NaOH), and aluminium oxide (Al_2O_3) with potassium permanganate ($KMnO_4$). To prevent humidity, dust, and other factors from damaging the MW generator's, a teflon window was installed between the isolator outlet and the MW cavity's inlet. The moisture content of the sludge samples, as well as the energy consumption of the system were monitored in real-time. The changes in the mass of the sludge while irradiated (water being evaporated) was measured by a single point load cell (Mettler Toledo). The energy supplied to the pilot system was measured using a power network analyser. Kocbek et al. (2020) reported a 72% conversion efficiency of the electrical energy to the MW energy at an MW power output of 6 kW. A more detailed description of the MW pilot-system can be found in Kocbek et al. (2020).

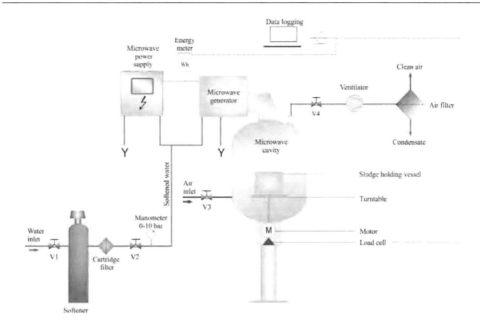

Figure 26: Schematic representation of the experimental pilot-scale MW system (Kocbek et al., 2020)

4.2.4 Experimental procedure

4.2.4.1 MW drying tests

After determining the initial moisture content of the sludge samples, three kg of each sludge sample were placed in the holding vessel of the MW system at a thickness of 60 ± 5 mm. The holding vessel containing the sludge sample was then placed on the rotating table and irradiated at an MW power output of 6 kW at a frequency of 2,450 MHz. All the experiments were performed either in duplicates or triplicates. The sludge samples were irradiated until reaching a moisture content of 0.18 kg of water kg of dry solids^{-1} (i.e., 85% DS). The experimental conditions for the evaluated samples are summarized in Table 5. The experimental work was carried out at the research hall of the Ptuj's municipal wastewater treatment plant (Ptuj, Slovenia).

Table 5: Experimental design

Sludge samples	WWTP location	Experimental parameters and variables			
		MW output power	Sludge mass	Sludge layer thickness	Final DS content
		[kW]	[kg]	[mm]	[kg of water kg of dry solids^{-1}]
WAS sample A	Ptuj	6.00	3.0	60 ± 5	0.18
WAS sample B	Ljutomer	6.00	3.0	60 ± 5	0.18
WAS sample C	Maribor	6.00	3.0	60 ± 5	0.18
SS sample D	Ptuj	6.00	3.0	60 ± 5	0.18

4.2.5 Data analysis

4.2.5.1 Specific energy input (SEI)

The SEI was calculated as shown in Equation 22:

$$SEI = \frac{P_{in;elect} \cdot t}{m_{sample}} \tag{22}$$

where the SEI is the specific energy input [kJ kg^{-1}], m_{sample} is the initial mass of sludge (kg), t is the exposure time [s], and $P_{in;elect}$ is the input power consumed by the system during the drying process [kW].

4.2.5.2 Energy efficiency (μ$_{en}$)

The μ$_{en}$ is the ratio between the theoretical energy demand for evaporating the water and the energy consumed by the MW unit during the drying process, the calculation for which is shown in Equation 23 (Jafari et al., 2018):

$$\mu_{en} = \frac{(m_{sample} \cdot c_p \cdot \Delta T) + (m_{eva} \cdot h_{fg})}{P_{in;elect} \cdot t} \cdot 100 \tag{23}$$

where μ$_{en}$ is the energy efficiency [%], c_p is the specific heat capacity of the water [J kg^{-1} °C^{-1}], ΔT is the temperature difference of the sample between the exposure time t and the start of the treatment, and h_{fg} is the latent heat of the water [J kg^{-1}] (2,257 x 10^3 J kg^{-1} at 100 °C as reported by Haque (1999)). The temperature of the samples was not measured; as such, the influence of sensible heat on total energy efficiency was considered assuming the sludge sample reached a temperature of 100 °C before water evaporation took place.

4.2.5.3 Sludge moisture content (X)

The X was calculated as shown in Equation 24 (Chen et al., 2014):

$$X = \frac{m_t - m_d}{m_d} \tag{24}$$

where X is the moisture content of the sludge [kg of water per kg of dry solid^{-1}], m_s is the sample mass at time t and m_d is the total mass of dry solids in the sample [kg]. The total mass of dry solids in the samples was determined as described in Section 4.2.2.1. The m_t was continuously determined by the point load cell located in the MW irradiation cavity, as described in Section 4.2.3. Therefore, X was continuously measured.

4.2.5.4 Drying rate (D$_R$)

The D$_R$ was calculated as shown in Equation 25 (Chen et al., 2014):

$$D_R = \frac{dX}{dt} \tag{25}$$

where D$_R$ is the drying rate [kg of water per kg of dry solid^{-1} min^{-1}] and X is the moisture content of the sludge at a specific exposure time. The drying rates were determined by polynomial regression analysis using Microsoft Excel and considered at a mass interval of 20 g.

4.2.5.5 Power absorption density

According to Maxwell equations, the power absorption density (i.e., the amount of power absorbed by a material per unit of volume) (P$_d$) is proportional to the input power, which relates to the electric field intensity (E), as shown in Equation 26 (Stuerga, 2006); Gupta et al. (2007):

$$P_d = 2\pi f \varepsilon_0 \varepsilon' |E|^2 \tag{26}$$

where P$_d$ is the amount of absorbed power per unit volume [Wm^{-3}], f is the microwave frequency [s^{-1}], ε_0 is the permittivity of free space [8.85 × 10^{-12} Fm^{-1}] and E is the electric field intensity [Vm^{-1}]. The electric field intensity can be calculated as described in Equation 27 (Soltysiak et al., 2008; Pitchai et al., 2012).

$$E = \sqrt{\frac{2P_{out,micr}}{1 - |S_{11}|^2}} \tag{27}$$

where P$_{out;micr}$ is the output power (nominal power) supplied to the MW chamber [kW] and, S$_{11}$ is the reflection coefficient associated with the fraction of the power reflected by the sample.

Assuming that the absorbed MW energy is converted into heat, the heating rate during the MW drying process can be related to power absorption density as follows (Clark et al., 2000; Beneroso et al., 2017):

$$\beta = \frac{P_d}{\rho c_p} \tag{28}$$

where β [°Cs^{-1}] is the heating rate (i.e., the temperature variation of the material with time) and ρ is the material density [kg m^{-3}].

4.3 Results and discussion

This section discussed the MW treatment performance when drying different types of sludge. Initially, the sludge samples were characterized to better understand the potential impact of the physical-chemical properties of the sludge on the MW treatment performance. Then, the MW treatment performance was assessed, determining the exposure time required to achieve a certain drying level, the sludge drying rates, energy efficiencies, and the specific energy inputs. The sludge samples were dried from their initial moisture content up to a final moisture content of 0.18 kg of water kg of dry solids^{-1} (85% DS).

4.3.1 Sludge physical-chemical characteristics

The physical-chemical characteristics of the evaluated sludge from the different sources and treatment facilities are depicted in Table 6. The average initial moisture content of the sludge samples evaluated in this study ranged between 79.6% and 84.3% (i.e., between 20.4 and 15.7% DS). The moisture content of the WAS samples A, B and C was on average 81 ± 1.5% (or 19 ± 1.5% DS). That reflects the good performance exhibited by the mechanical dewatering unit (i.e., centrifuge) on the sludge dewatering. Similar values were reported in the literature, with sludge moisture content ranging between 79 to 87% (13 to 21% DS) (Léonard et al., 2004; Mawioo et al., 2017). The effectiveness of the conditioning methods for dewatering the SS samples were also effective as reflected in the SS moisture content of 84.3% (15.7% DS). The SS (sample D) samples were flocculated, dewatered by gravity, and sieved.

The organic content of the evaluated samples (expressed as VS) shown in Table 6 ranged between 68 and 88%. Such variability could be eventually explained due to either the inherent properties of the different types of evaluated sludge, or to the treatment processes the sludge went through. The VS concentrations of the WWTP A, B, and C were similar with an average VS concentration of 86 ± 4 %; these values were in accordance with : (i) the existent wastewater treatment processes at the WWTPs, and (ii) with the absence of sludge stabilization processes at the WWTPs such as anaerobic or aerobic sludge digestion(Léonard et al., 2004; Mawioo et al., 2017; Akdağ et al., 2018). The lowest VS fraction was obtained for the sludge sample D at a VS concentration of 68%. This value could be eventually attributed to the long sludge holding time of the sludge (five years or longer) at the storage SS tank in the WWTP; thus, the degradation of some of the organic material could occur. Considering such long retention time at which the SS was exposed and the subsequent reduction in the VS concentration, it was also expected a potential reduction on the gross calorific value of such SS. However, as shown in Table 6, the gross and net calorific values of the SS (sample D) were comparable with the values obtained for the WAS samples A, B, and C. Average values for the gross and net calorific values of 19 ± 0.5 MJ kg^{-1} and 17.3 ± 0.6 MJ kg^{-1}, respectively were obtained for the evaluated samples. Therefore, there were no major variations on both the gross and net calorific values considering the large variations on the VS concentrations. In addition, the C, H, N, and S content of the different samples (WAS vs SS) did not change considerably. Average concentrations for C, H, N, and S of 43.6 ± 3.2%, 5.3 ± 0.4%, 7.2 ± 3.1%, and 1.5 ± 0.1%, respectively were reported for all the evaluated samples. The high calorific value obtained for the SS (sample D), could be then attributed to the higher oil and fat content reported for the SS samples compared to the WAS samples (1.36 g gDS^{-1} for the SS compared to a range from 0.28 to 0.35 g gDS^{-1} for the WAS sludge samples). The SS samples were obtained from septic tanks receiving household wastes that could have included waste cooking vegetable oil, and similar products with a high oil and grease content. As such, the high oil and grease content on the SS samples could have exhibited a positive effect on the SS samples regarding their energetic value. For instance, the amount of heat released during the combustion of waste cooking vegetable oil is higher than 40 MJ kg^{-1} (Fassinou, 2012). So, when vegetable oils were present in the sludge matrix, the calorific values observed in sludge samples could increase.

The sludge floc structure is usually porous and contains voids through which the fluids (i.e., water) can move through (Cui et al., 2019). Porosity was calculated by dividing the difference between the envelope and absolute volume by the envelope volume. The porosity values for the evaluated samples are reported in Table 6. The porosity of the evaluated sludge samples ranged between 43 and 64%. Higher porosity values were observed for the SS (sample D) than for the WAS sludge samples. That could be eventually attributed to the large retention time at which the SS was retained in the septic tanks, eventually promoting the degradation of sludge. Following the Kozeny – Carman equations, the higher the porosity, the higher the permeability of the sludge (Schulz et al., 2019); thus, the higher the potential dewaterability of the sludge. Therefore, the SS (D) could eventually exhibit a higher tendency for losing water compared to the other sludge samples. However, other factors should have been also considered such as the particle size distribution, the size and shape of the pores, as well as their level of connectedness, which were not evaluated in this study (Schulz et al., 2019).

The concentrations of metals present in the sludge samples including Ni, Zn, Pb, Cr, and Cu complied with the European standards for agricultural reuse of 400 mg kg^{-1} for Ni, 4,000 mg kg^{-1} for Zn, 1,200 mg kg^{-1} for Pb, and 1,750 mg kg^{-1} for Cu (Wiśniowska et al., 2019). However, most European countries adopted more stringent regulations with respect to land applications of treated sludge. For instance, the Slovenian regulation for sludge reuse in agriculture set the following standards: 30 mg kg^{-1} for Ni, 100 mg kg^{-1} for Zn, 40 mg kg^{-1} for Pb, 40 mg kg^{-1} for Cr, and 0.5 mg kg^{-1} for Cu (Wiśniowska et al., 2019). These results indicated that the potential agricultural reuse of the treated sludge derived from WWTP and SS tanks would significantly vary depending on the local regulations. Nevertheless, sludge not suitable for agriculture, can be alternatively converted into stable fractions through (co-) combustion, while generating heat and electricity.

Table 6: Physical-chemical characteristics of the evaluated sludge samples

Parameter	Unit	Sample A	Sample B	Sample C	Sample D
Moisture content	[%]	83 ± 1.0	79.6 ± 0.2	80.0 ± 0.1	84.3 ± 0.2
Dry solids[a]	[%]	17 ± 1.0	20.4 ± 0.2	20.0 ± 0.1	15.7 ± 0.2
Volatile solids	[%]	88 ± 2	81 ± 0.2	88 ± 0.3	68 ± 1.2
C[a]	[%]	46.5	39.8	46.1	42.0
H[a]	[%]	5.0	5.5	5.7	4.8
N[a]	[%]	9.9	9.3	6.51	3.2
S[a]	[%]	1.5	1.5	1.3	1.5
Gross calorific value	[MJ kg^{-1}]	18.4	19.0	19.6	18.9
Net calorific value[a]	[MJ kg^{-1}]	16.7	17.3	17.9	17.2
Oil and grease	[g gTS^{-1}]	0.35 ± 0.01	0.31 ± 0.003	0.28 ± 0.16	1.36 ± 0.04
Absolute density[a]	[g cm^{-3}]	(141 ± 0.13) x 10^{-2}	(144 ± 0.05) x 10^{-2}	(141 ± 0.04) x 10^{-2}	(149 ± 0.05) x 10^{-2}
Absolute volume[a]	[cm^3]	(405 ± 0.36) x 10^{-2}	(342 ± 0.11) x 10^{-2}	(284 ± 0.09) x 10^{-2}	(178 ± 0.05) x 10^{-2}
Envelope density[a]	[g cm^{-3}]	(80 ± 0.50) x 10^{-2}	(80 ± 0.50) x 10^{-2}	(71 ± 0.40) x 10^{-2}	(54 ± 0.50) x 10^{-2}
Envelope volume[a]	[cm^3]	7.16	6.13	5.67	4.97
Porosity[a]	[%]	43	44	50	64
Nickel (Ni)[a]	[mg kg^{-1}]	23.2	57.9	19.7	22.4
Zinc (Zn)[a]	[mg kg^{-1}]	437	1403	524	872
Lead (Pb)[a]	[mg kg^{-1}]	25	71	29	45
Chromium (Cr)[a]	[mg kg^{-1}]	68	2981	51	82
Copper (Cu)[a]	[mg kg^{-1}]	55	70	75	61

[a] dry basis

Figure 27 depicts the adsorption/desorption behaviour of the MW dried sludge samples (moisture sorption isotherms). For all the evaluated samples, similar shapes on the sorption isotherms were obtained with an initial low water sorption/desorption at low RHs, and a substantial increase at high RHs (above 75%). The curves indicates a an S-shape, commonly observed in organic materials, such as sludge (Vaxelaire et al., 2001; Freire et al., 2007) and provides an insight into different water-binding mechanisms at the individual sites of the sludge (Mujumdar et al., 2000). Several models have been proposed to explain the shape of the S-shape sorption isotherm shown in Figure 27. Hailwood et al. (1946), suggested that water adsorbed onto the sludge exist in two forms: (i) water of hydration, corresponding to the water molecules bound to the OH groups in the sorbent material (sludge in this case) by polar interactions (monolayer water); and (ii) solid solution or dissolved water, corresponding to the water molecules slightly bound to the sorbent material, but still located within the porous structure of the sludge (poly-layer water). As a result, the strength of the water molecules bound to sludge during the sorption/desorption processes would depend on the presence and availability of hydrophilic and polar groups responsible for producing strong intermolecular interactions between the water molecules and the sludge as described in the type (i) water-sludge interaction (water of hydration or monolayer water). From looking at Figure 27, three different regions can be distinguished, which may eventually indicate different binding characteristics between the water molecules and the sludge. The first region can be observed at low RH (left side of Figure 27 towards the first bend, from 0 to approximately 20%). In that region, it seems that it is more difficult for the water molecules to get adsorbed/desorbed in the sludge matrix. The water molecules may interact with the hydrophilic groups of the material such as the OH or NH groups (Peleg, 2020). This was described as the type (i) interaction and/or monolayer adsorbed water. Thus, the lower part of the isotherm is characterized by water molecules tightly bound to the sludge. The enthalpy of evaporation for that type of water as described in that region has been reported by other authors to be much higher than that of pure water confirming such strong bonding between the water molecules and the sludge (Andrade et al., 2011). In the second region (between the first (20%) and second bends (75%), the slope of the curve is higher than in the first region and the water molecules seem to be less tightly bound to the sludge compared to the first region. In this region eventually different layers of water could form one on top of each other (that is, on top of the first monolayers) (Bougayr et al., 2018). The enthalpy of evaporation for the water molecules in this zone has been reported by other authors to be slightly higher than that of pure water (Penfield et al., 1990; Andrade et al., 2011). In the third region (above the second bend, above 75%), the slope becomes steeper and the water molecules seemed to be loosely bound; mainly the water molecules can be adsorbed in cavities in the sludge matrix, in large capillaries, and in sludge voids (Penfield et al., 1990; Andrade et al., 2011). As the RH increased, the sludge cavities and voids seemed to become increasingly occupied. Then, the absorption of the water molecules could occur onto less active sites involving lower interaction energies. When the RH approaches 100%, the curve leans to a vertical asymptote, which represents the free water present in the sludge (Bougayr et al., 2018). In conclusion, the intermolecular forces between the water molecules and the sludge observed at the lower RHs (characterized by the presence of adsorption active sites in the sludge such as OH or NH groups - monolayer adsorption) were much stronger that the forces involved in the interaction between the water molecules and the sludge at higher RHs (characterized by the water bound to the sludge cavities and voids - multilayer adsorption). Correspondingly, the amount of energy that must be supplied to the material during desorption (drying) would strongly depend on the characteristics of the molecular bonding between the water and the sludge.

As observed in Figure 27, the different sludge samples showed different affinities for adsorbing/desorbing water at the evaluated RHs. The SS (sample D) showed the lowest EMCs at the evaluated RHs. The WAS samples (samples A, B, and C) exhibited similar EMCs at the entire evaluated RHs, slightly higher than for the SS (sample D). For instance, at a 95% RH (Figure 27a), the SS (sample D) reached the lowest EMC value of 0.15 kg of water kg of dry solids^{-1} (or 87% DS), compared to the WAS sample A (0.35 kg of water kg of dry solids^{-1} or 74% DS), WAS sample C (0.38 kg of water kg of dry solids^{-1} or 72% DS), and WAS sample B (0.43 kg of water kg of dry solids^{-1} or 70% DS). The SS (sample D) was characterized by the lowest amount of moisture content (at 95% RH); so, the less amount of bound water. This could be eventually related to the organic and oil and grease content in the sludge; these values were presented in Table 6. The SS (sample D) contained a much lower concentration of organic matter compared to the WAS samples; thus, less available sites for the water molecules to get adsorbed. In addition, oil and grease compounds are highly hydrophobic, having a significant effect on the uptake of water from the environment; consequently, resulting in low EMC values. As indicated both in Figure 27 and Table 6, there is an inverse relation between the EMC and the oil and grease content in the sludge; that is, the higher the oil and grease concentration in the sludge, the lower the EMC values. Thus, the presence of oil and grease together with a low organic matter content seemed to increase the hydrophobicity of the sludge. Having such high hydrophobicity is a desirable property of a material when the goal is to dry such material.

Therefore, sludge samples with such high hydrophobic (as SS – sample D) would eventually exhibit a better performance when exposed to MW drying.

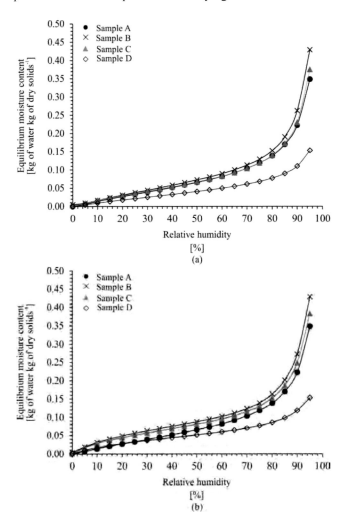

Figure 27: Equilibrium moisture content (EMC) as a function of the RH (moisture sorption/desorption isotherms) for the evaluated sludge samples during (a) adsorption and (b) desorption.

The isosteric heat of sorption provides an indication of the strength of the interaction between the water molecules and the material. In this study, the isosteric heat of sorption was assessed for the WAS sample A at a sludge moisture content ranging from 0.01 to 0.38 kg of water kg of dry solids[-1] (i.e., from 1.4 to 72% DS content) following the same procedure as for obtaining the moisture sorption isotherms described in Figure 27. As shown in Figure 28, the isosteric heat of sorption decreased as the sludge moisture content increased in the evaluated sludge moisture content range; in addition, the isosteric heat of sorption at the entire sludge moisture content evaluated range was always higher than the heat of evaporation of pure water (2.4 MJ kg^{-1}). These results clearly indicate that the lower the moisture content in the sludge, the higher the energy required to remove that water from the sludge. That is, the lower the moisture content the higher the presence of water molecules bound to the sludge to polar sludge active sites such as the OH and NH groups (monolayer water), so a higher amount of energy would be required to break such water bonding. As the sludge moisture content increased, the energy required to remove the water out of the sludge decreased. Therefore, it is expected that the energy required by a MW system for drying such type of sludge would increase as the sludge moisture content decreases. In this study the isosteric heat of sorption was only determined for the WAS sample A, but it was not determined for the other sludge samples. However, looking at the moisture sorption/desorption isotherms presented in Figure 27, no majors differences on the isosteric heat of sorption are expected between all the evaluated sludge samples; eventually, the SS sample D would exhibit a little lower isosteric heat of sorption compared to the other sludge samples.

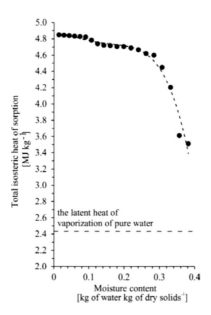

Figure 28: Isosteric heat of sorption as a function of moisture content

4.3.2 Sludge MW drying: drying rate performance

The sludge samples were irradiated at the experimental conditions described in Table 5. Figure 29 shows the changes in the moisture content of the evaluated sludge samples as a function of the exposure times. Figure 30 indicates the changes in the drying rates of the sludge samples as a function of the exposure time (Figure 30a) and moisture content (Figure 30b). Similar shapes were observed for the four evaluated sludge samples; however, sludge samples D and A clearly contained a higher initial moisture content. Three different drying periods can be clearly observed in Figure 30 as follows: (i) the adaptation drying period, (ii) the constant rate drying period, and (iii) the falling rate drying period. At an early stage of the drying process in the adaptation drying period, the temperature of the sludge rises until the vapour pressure of the evaporated water is equal to the surrounding pressure. The water in the sludge sample evaporated at the atmospheric boiling point (100 °C) (Doran, 2013; Berk, 2018). As the temperature increased, the drying rates also increased as observed both on the left-hand side of Figure 30a and on the right-hand side of Figure 30b. The maximum drying rate indicated the end of the adaptation phase and the beginning of the constant rate drying period. The length of the adaptation phase ranged from 7 to 8 minutes for all the evaluated samples, depending on the initial moisture content of the sludge samples. Notably, the higher the initial moisture content of the sludge samples (Figure 29), the higher the maximum drying rates observed in Figure 30. A reduction in the moisture content of the sludge samples of approximately 23% (1.1 kg of water kg of dry solids^{-1}) was observed in the adaptation period. Following the initial adaptation drying rate period, a steady state drying rate was achieved. During that period, the water evaporated consisted of the free water present on the surface of the sludge. The constant drying period lasted as long as the water transportation rate from the inner part of the sludge to the surface was equal or higher than the water evaporation rate from the surface of the sludge (Doran, 2013; Berk, 2018). The constant rate drying period lasted for a considerably large amount of time, almost until the drying process was finished. This observation is in agreement with previous studies reported in the literature (Dominguez et al., 2004; Chen et al., 2014; Bennamoun et al., 2016; Mawioo et al., 2017; Kocbek et al., 2020). Dominguez et al. (2004) reported that MW dryers extended the duration of the constant rate drying period compared to traditional conductive and convective dryers. Ni et al. (1999) and Fu et al. (2017) observed that the increase in the constant drying rate periods (i.e. removal of free and loosely bound water) could be attributed to the selective penetration of the MW irradiation to certain absorbent materials; therefore, causing heat generation from the inside of the target material creating an inverted temperature profile compared to conventional dryers (i.e. in MW irradiation the temperature is higher inside the material rather than in the surface of the material). This heat, a result of molecular friction, causes internal evaporation which in turn increases the internal pressure promoting the water mass transfer to the surface of the material. Such process accelerates the drying rates and extends the constant drying rate period in MW drying. As the drying process continued, the sludge moisture content of the sludge samples further decreased, and the drying rates started to fall. At that point, the process reached the falling drying rate period. At the start of the falling rate period, the amount of the free water on the surface of the sludge declined; the liquid film on the surface of the solid is no longer present (Doran, 2013; Berk, 2018). This condition is gradually expanded until the surface of the material is completely dry. So, the drying process continued, but the evaporating surface moves into inside the material (Doran, 2013; Berk, 2018) becoming harder than before to transfer the moisture from the capillaries and interstices of the material to the surface (Ma et al., 2017). As such, the drying rate decreased since the water removal process was now being governed by internal diffusion mechanisms (much slower than the evaporation of the free water). The start of such falling drying rate period also indicated the end of the free water removal, and the start of the bound water removal. An interesting observation from looking at the results in Figure 30, is that the drying falling rate period was attained faster in sludge samples from the centralized WWTP (A,

B and C) compared to the sludge sample taken from the septic tanks (D); thus, confirming that sludge sample D contained less bound water compared to WAS samples (as discussed in Section 4.3.1). In conclusion, the concentration of organic matter in the sludge, as well as the presence of oil and grease have an impact on the MW drying performance of the sludge. The results presented in this section validated that MW irradiation accelerated the evaporation of water from the surface of the material, while extending the duration of the constant drying rate period. This allowed the unbound water to be removed regardless of the type of sludge being evaluated.

In addition, Ma et al. (2017) and Antunes et al. (2018) reported that the rate of the water evaporation (from the sludge surface when applying MW radiation) was influenced by the rate at which the electromagnetic energy is absorbed by the material and converted into heat (power absorption density (P_d)). This has been reported to largely depend on several factors including both the electric field strength (which depends on the MW power output and sample volume), as well as the dielectric properties of the sludge such as the dielectric loss factor (ε'') as described in Equation 26; particularly, the sludge moisture content directly impact on the dielectric loss factor (ε'') (Ma et al., 2017; Antunes et al., 2018). For instance, Ma et al. (2017) reported dielectric loss factor (ε'') of 14.1 for municipal dewatered sludge at a moisture content of approximately 81% (19% DS). As the sludge was dried (i.e. as the water content of the sludge was reduced) up to a sludge moisture content of 6% (94% DS), a the dielectric loss factor (ε'') of 0.3 was reported (Ma et al., 2017). Similar values were also reported by Antunes et al. (2018). That is, the moisture content of the sludge has a direct impact on the dielectric loss factor; therefore, on the power absorption density (P_d – as described in Equation 26), and on the rate at which the MW electromagnetic energy is absorbed into the sludge samples. The water molecules strongly interact with the applied MW field. So, the higher the moisture content (i.e., the higher the power density), the higher also the heating rate β as described in Equation 28; that is, the higher the rate at which the temperature of the sludge increases. The higher the heating rates, the higher the internal vapour pressure gradient. Therefore, leading to an increase in mass transfer of the water to the surface, and so to the drying rates (Kumar et al., 2016). Such trends were clearly observed in this study. The sludge sample with the highest moisture content (SS sample D – Table 6) exhibited the higher drying rates (Figure 30). Such higher drying rates observed for the sludge samples with a higher moisture content, allowed the same exposure time required for all the samples to reach a final moisture content to 0.18 kg of water kg of dry solids^{-1} (85% DS content) as shown in Figure 29, regardless the initial moisture content of the water. The exposure time needed to dry the sample up to 0.18 kg of water kg of dry solids^{-1} (85% DS content) was similar across the entire range of evaluated sludge samples at an average of 21 ± 0.4 minutes (Figure 29 and Figure 30). Or expressed in an opposite mean, the favourable conditions for the absorption and conversion of MW energy into heat observed at a higher sludge moisture content was counteracted by the higher net amount of water initially present in the sludge sample that needed to be removed to reach the desired drying goal. Therefore, the increase in the dielectric properties of the irradiated material does not necessarily mean lower exposure time for achieving a desired level of moisture content.

The water molecules can be present in the sludge either: (i) free or loosely attached to the sludge, or (ii) tightly bound to the sludge (Vesilind, 1994). The bound water is more strongly attached to the sludge compared to the free water or loosely attached water. According to Jones et al. (2003), both the free and loosely attached water molecules to the material exhibit low physical attractions to the material and can move more freely compared to the bound water. Hence, such water molecules can be easily rotated under the oscillating electromagnetic field provided by the MW irradiation; so, resulting in friction and heat. On the other hand, the effective movement of the bound water molecules is constrained by the strong physical binding forces between the solid material and the water molecules. Consequently, those water molecules cannot effectively absorb and convert the energy generated by the kinetic movements into heat; therefore, resulting in lower drying rates (Jones et al., 2003). Such water fractions of the sludge (free and bound) can vary considerably depending on the source of the wastewater and the treatment processes at which the sludge was subjected to. The amount of bound water present in the sludge was determined in this study by assessing the water sorption isotherms of the MW dried sludge (shown in Section 4.3.1). The SS sludge (sample D) exhibited the lowest amount of bound water in the sludge compared to the WAS samples (A, B and C). Thus, the SS (sample D) (exhibiting an initial high water moisture content) could have absorbed the MW energy more efficiently leading to the higher drying rates shown in Figure 29 and Figure 30. In addition, due to the higher free water content of the SS (sample D), longer constant drying periods were observed for these samples as shown in Figure 29 and Figure 30 compared to the WAS samples. Therefore, the results presented in this section demonstrated that the quantity and the distribution of water within the sludge matrix have a substantial impact on the MW sludge drying performance by affecting the absorption and conversion of MW energy into heat. Specifically, the sludge samples characterized with a higher amount of free and loosely bound water promoted a higher MW energy absorption and conversion of energy into heat leading to higher drying rates and to the extension of the constant drying rate duration.

Furthermore, the higher the oil and grease content of the sludge, the higher the amount of free or loosely bound water in the sludge; that is, the higher the hydrophobicity of sludge. The presence of oil and grease materials in the sludge (e.g., fats and oils) could also positively impact the thermal conductivity and specific heat capacity of the sludge (Lyng et al., 2014). Oil and grease materials exhibit higher thermal conductivities and lower specific heat capacities than water; thus, requiring less energy for rising the temperature of the heated material (Lyng et al., 2014). Therefore, the presence of oil and grease components in the SS (sample D) (showing the highest oil and grease concentrations compared to the other sludge samples – Table 6) could also contribute to the high drying rates observed for the SS (sample D) compared to the other sludge samples (Figure 29 and Figure 30). Other compounds present in the sludge samples, such as the organic matter content could also influence the sludge drying process. Likewise, other physical-chemical properties such as the porosity and permeability could promote faster MW drying rates by providing better possibilities for the water molecules and water vapour to diffuse out of the material (Pickles et al., 2014). The SS sample (D) as shown in Table 6 also exhibited the highest porosity values compared to the other evaluated sludge, so that could probably also contribute to the highest drying rate observed for the SS. Therefore, the MW drying performance strongly depend on the sludge physical-chemical properties determined and presented in this study such as the moisture content, the porosity, the organic matter content (VS content), the oil and grease content, and the type of interaction between the water molecules and the sludge, among others. These properties exhibited a strong influence on the MW drying performance for the treatment of sludge, and the better these properties are known, the better the performance of the MW drying system can be predicted.

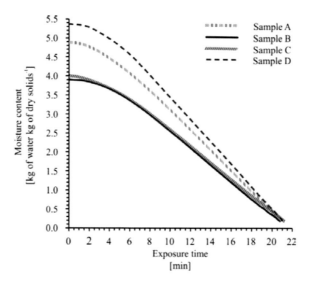

Figure 29: Moisture content as a function of the exposure time for the evaluated sludge samples.

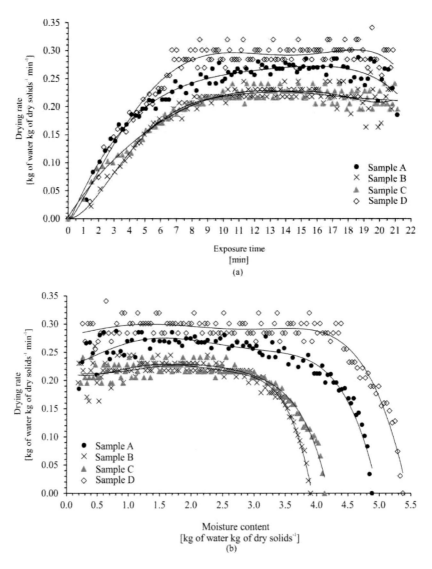

Figure 30: Drying rates as a function of the (a) exposure time, and (b) moisture content

Figure 31: WAS sample A; from left to right; mechanically dewatered WAS, dried WAS sample

Figure 32: SS sample D; from left to right; raw SS, dewatered SS, MW dried SS sample, SS filtrate.

Figure 33: MW dried sludge samples

4.3.3 Energy performance of the system

The energy performance of the MW system was also evaluated in terms of both the energy efficiency μ_{en} (Equation 23), and the actual MW energy consumed for drying the sludge samples until the desired (final) moisture content (defined as the SEI as described in Equation 22). One alternative for measuring the energy efficiency performance of a drying system is to look at the ratio between the latent heat of the water and the actual energy consumed during the drying process (Earle et al., 2004). However, as described in Equation 23, in this study the sensible heat was also factored (in addition to the latent heat) as it can be possible to recover that energy as heat. Both the energy efficiency (μ_{en}) as well as the SEI calculations were based on the actual electrical power consumed by the MW system during the drying process. Therefore, such calculations (indicated by Equations 22 and 23) strongly depended on the conversion efficiency of the delivered electrical power to the MW energy (i.e., on the MW generation efficiency) (Lakshmi et al., 2007; Jang et al., 2011). Such MW generation efficiency was previously reported by Kocbek et al. (2020) for the same MW system at 72% when working at a MW power output of 6 kW (as in this study).

Figure 34 and Figure 35 illustrate the SEI and energy efficiency as a function of the sludge moisture content for all the evaluated samples. All the sludge samples were dried up to a final moisture content of 0.18 kg of water kg of dry solids^{-1} (85% DS). From looking at Figure 34, it can be observed that the lower the initial moisture content of the sludge sample (for instance sample B vs sample D), the lower the SEI required to reach certain degree of dryness; therefore, the higher the energy efficiency to reach that particular level of moisture content (or DS content). Both figures closely follow the trends previously observed regarding the changes in the sludge moisture content and drying rates as presented in Section 4.3.2. For instance, at the start of the drying process, an adaptation drying rate period can be observed. In this period, the energy was used for increasing the temperature of the sludge. The energy efficiency values reported in Figure 35 were calculated following the Equation 23. Such an equation includes in the nominator both the latent heat, and the heat capacity of the water. So, at the early stages of the drying process (when the drying process was started) the sludge sample was mostly being heated without any or little water being evaporated. So, the theoretical energy demand (as expressed in the numerator of Equation 23) mostly considered the heat capacity of heating the water, which is relatively small compared to the latent heat of evaporation (since little or no water was evaporated at that time). Therefore, that efficiency term shown in Figure 35 exhibited low values corresponding to that adaptation drying phase. Given that most of the MW energy was used to increase the temperature of the sludge, only small amounts of water were actually removed. Following the adaptation phase period, there were marginal changes both in the changes of the SEI as a function of the moisture content, as well as in the and energy efficiencies as a function of the moisture content for all the evaluated samples (Figure 34 and Figure 35). This could be explained considering that the system was removing the surface water of the material at steady-state conditions (i.e., in the constant drying rate period). Looking from an energy performance standpoint, it was not possible to observe the final falling rate drying period in Figure 34 and Figure 35 as observed in Figure 30. A similar situation was observed and reported by Kocbek et al. (2020).

The energy performance of the MW system seemed to vary depending on the sludge composition. For instance, when drying the different sludge samples up to a moisture content of 2.5 kg of water kg of dry solids^{-1} (i.e., up to 50% DS), the SEI required for the SS sample (D), as shown in Figure 34, was 2.2 MJ kg^{-1} (0.6 kWh kg^{-1}) compared to the SEI required for WAS sample B of 1.7 MJ kg^{-1} (0.5 kWh kg^{-1}). These differences become less pronounced as the water is being removed from the sludge samples. For instance, at the end of the drying process the SEIs were similar across the entire range of evaluated samples averaging 3.5 ± 0.07 MJ kg^{-1} (1.0 ± 0.02 kWh kg^{-1}). As described in Section 4.3.2, that behaviour could be related to both the efficiency of the MW absorption, as well as to the conversion of the electromagnetic energy into heat mostly governed by the dielectric properties of the material. These properties strongly depend both on the sludge moisture content, and on the distribution of the water fractions within the material (i.e., the free and bound water content). For instance, sludge samples with a higher amount of free water exhibited a high energy absorption rate within the material; thus, large sludge drying rates (Figure 30). This could also lead to the extended constant drying rate period associated with the removal of unbound water as observed in Figure 30. In turns, as explained in Section 4.3.2, higher sludge drying rates were observed in those sludge samples having a higher proportion of free water than bound water. When particularly focusing on the energy requirement and overall energy efficiency, Figure 34 and Figure 35 show that the SS sample (D), characterized with the highest initial water content (and also free water content) from all the evaluated samples, required initially the highest SEI to evaporated the initial amounts of water; however, the SEI requirements to completely dry the material up to 0.18 kg of water kg of dry solids^{-1} (85% DS) was the same as for the other evaluated samples. When looking at the energy efficiency (Figure 35), it was observed that the energy efficiency for the SS (sample D) also reached higher energy efficiency values at higher moisture contents compared to the other evaluated sludge samples. For instance, the SS (sample D) reached energy efficiency values higher than 60% at moisture contents as high as 4.5 kg of water kg of dry solids^{-1} compared to the rest of the evaluated sludge samples that reached that level of energy efficiency somewhere between 3 and 4 kg of water kg of dry solids^{-1}.

The sludge physical-chemical properties exhibited an important role in determining the MW energy absorption performance, which may explain the overall energy efficiency performance and SEI requirements. The MW energy absorption and conversion into heat increased with the amount of free and loosely bound water fraction present in the sludge samples. This had an effect, extending the duration of the constant drying period phase, leading to an overall increase in the MW energy efficiency. Thus, the higher the concentration of free water in the sludge, the more energy-efficient MW drying process (both heating the sludge sample and evaporating the water out of the sludge). The amount of free and loosely bound water molecules was found to be dependent on several physical-chemical properties including the sludge organic content and the concentration of hydrophobic compounds present in the sludge such as the presence of oils and grease. The oil and grease content could also have a positive impact both on the energy needs to increase the temperature of the sludge, as well as on contributing to a more uniform temperature distribution within the bulk material. Similarly, the sludge porosity could positively impact the water flow out of the sludge, improving the MW drying and energy performance. Therefore, these results indicate that the origin of sludge, as well as the wastewater treatment and stabilization process at which that sludge has been exposed, have a large influence on the physical-chemical characteristics of sludge and as such on the MW system performance. Examining such properties may contribute to better understanding and optimising the MW sludge drying performance.

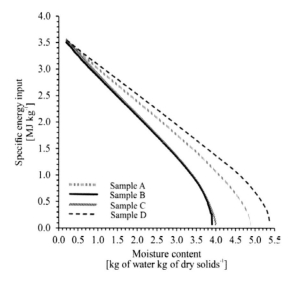

Figure 34: Specific energy input as a function of moisture content for the evaluated sludge samples

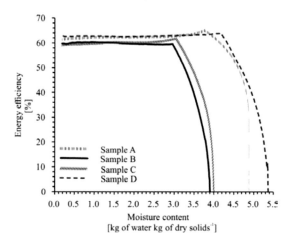

Figure 35: Energy efficiency as a function of moisture content for the evaluated sludge samples

4.4 Conclusions

- Both the origin of the sludge (SS vs WAS), as well as the type of wastewater treatment process and biological stabilization processes at which the sludge was exposed impacted on the physical-chemical characteristics of the sludge including the moisture content, organic content, porosity, presence of hydrophobic compounds, and the water molecule-sludge interactions, among others.

- The MW drying extends the constant rate drying period, thereby accelerating the free water evaporation from the surface of the material which occurs almost independently of the type of sludge that was irradiated.

- The MW energy absorbed by the material on a given time interval is a critical parameter influencing the duration and intensity of the MW drying process; such energy absorption process depends on the dielectric properties of the materials.

- The MW absorption efficiency increased with an increase in the free water content of the sludge; thus, contributing to higher heating rates and increasing the system throughput treatment capacity.

- The MW drying performance depends on the physical-chemical properties of the sludge. Sludge containing a high fraction of free (unbound) water efficiently absorb and convert MW energy into heat

- The sludge unbound water content was higher in the sludge samples showing high concentrations of oil and grease compounds and low organic matter content. These conditions were favoured in septic tanks. Thus, the SS (sample D) absorbed and converted MW energy into heat more efficiently compared to the WAS samples derived from WWTP.

References

Akdağ AS, Atak O, Atimtay AT, Sanin FD. (2018). Co-combustion of sewage sludge from different treatment processes and a lignite coal in a laboratory scale combustor. Energy, 158, 417-426.

Andrade RD, Lemus R, Perez CEJV. (2011). Models of sorption isotherms for food: uses and limitations. 18(3), 325-334.

Antunes E, Jacob MV, Brodie G, Schneider PA. (2018). Microwave pyrolysis of sewage biosolids: Dielectric properties, microwave susceptor role and its impact on biochar properties. Journal of Analytical and Applied Pyrolysis, 129, 93-100.

Aviara NA. (2020). Moisture sorption isotherms and isotherm model performance evaluation for food and agricultural products. In Sorption in 2020s (pp. 145): IntechOpen.

Beneroso D, Monti T, Kostas ET, Robinson J. (2017). Microwave pyrolysis of biomass for bio-oil production: Scalable processing concepts. Chemical Engineering Journal, 316, 481-498.

Bennamoun L, Chen Z, Afzal MT. (2016). Microwave drying of wastewater sludge: Experimental and modeling study. Drying Technology, 34(2), 235-243.

Berk Z. (2018). Food process engineering and technology: Academic Press.

Bilecka I, Niederberger M. (2010). Microwave chemistry for inorganic nanomaterials synthesis. Nanoscale, 2(8), 1358-1374.

Bougayr EH, Lakhal EK, Idlimam A, Lamharrar A, Kouhila M, Berroug F. (2018). Experimental study of hygroscopic equilibrium and thermodynamic properties of sewage sludge. Applied Thermal Engineering, 143, 521-531.

Candanedo L, Derome D. (2005). Numerical simulation of water absorption in softwood. Paper presented at the Proceedings of the 9th International IBPSA Conference, August.

Chen Z, Afzal MT, Salema AA. (2014). Microwave drying of wastewater sewage sludge. Journal of Clean Energy Technologies, 2(3), 282-286.

Clark DE, Folz DC, West JK. (2000). Processing materials with microwave energy. Materials Science and Engineering: A, 287(2), 153-158.

Cui Y, Ravnik J, Steinmann P, Hriberšek M. (2019). Settling characteristics of nonspherical porous sludge flocs with nonhomogeneous mass distribution. Water Research, 158, 159-170.

Dealler SF, Rotowa NAL, Richard W. (1992). Ionized molecules reduce penetration of microwaves into food. International journal of food science & technology, 27(2), 153-157.

Dominguez A, Menéndez J, Inguanzo MP. (2004). Sewage sludge drying using microwave energy and characterization by IRTF. Afinidad, 61(512), 280-285.

Doran PM. (2013). Chapter 11 - Unit Operations. In Doran PM (Ed.), Bioprocess Engineering Principles (Second Edition) (pp. 445-595). London: Academic Press.

Đurđević D, Trstenjak M, Hulenić I. (2020). Sewage Sludge Thermal Treatment Technology Selection by Utilizing the Analytical Hierarchy Process. Water, 12(5), 1255.

Earle R, Earle M. (2004). Unit operations in food processing, web edition. The New Zealand.

Fassinou WF. (2012). Higher heating value (HHV) of vegetable oils, fats and biodiesels evaluation based on their pure fatty acids' HHV. Energy, 45(1), 798-805.

Flaga A. (2005). Sludge drying. Paper presented at the Proceedings of Polish-Swedish seminars, Integration and optimization of urban sanitation systems. Cracow March.

Freire F, Freire FB, Pires E, Freire JJEt. (2007). Moisture adsorption and desorption behavior of sludge powder. 28(11), 1195-1203.

Friedl A, Padouvas E, Rotter H, Varmuza K. (2005). Prediction of heating values of biomass fuel from elemental composition. Analytica Chimica Acta, 544(1), 191-198.

Fu B, Chen M, Song J. (2017). Investigation on the microwave drying kinetics and pumping phenomenon of lignite spheres. Applied Thermal Engineering, 124, 371-380.

Gomes LA, Gabriel N, Gando-Ferreira LM, Góis JC, Quina MJ. (2019). Analysis of potentially toxic metal constraints to apply sewage sludge in Portuguese agricultural soils. Environmental Science and Pollution Research, 26(25), 26000-26014.

Gupta M, Leong EWW. (2007). Microwaves and metals: John Wiley & Sons.

Hailwood A, Horrobin SJTotFS. (1946). Absorption of water by polymers: analysis in terms of a simple model. 42, B084-B092.

Haque KE. (1999). Microwave energy for mineral treatment processes—a brief review. International Journal of Mineral Processing, 57(1), 1-24.

Jafari H, Kalantari D, Azadbakht M. (2018). Energy consumption and qualitative evaluation of a continuous band microwave dryer for rice paddy drying. Energy, 142, 647-654.

Jang S-R, Ryoo H-J, Ahn S-H, Kim J, Rim GH. (2011). Development and optimization of high-voltage power supply system for industrial magnetron. Transactions on Industrial Electronics, 59(3), 1453-1461.

Jones SB, Or D. (2003). Modeled effects on permittivity measurements of water content in high surface area porous media. Physica B: Condensed Matter, 338(1), 284-290.

Kamran K, Gao N. (2020). Thermochemical Conversion of Sewage Sludge: A Critical Review. Progress in Energy and Combustion Science, 79.

Kehrein P, van Loosdrecht M, Osseweijer P, Garfí M, Dewulf J, Posada J. (2020). A critical review of resource recovery from municipal wastewater treatment plants–market supply potentials, technologies and bottlenecks. Environmental Science: Water Research & Technology, 6(4), 877-910.

Kelessidis A, Stasinakis AS. (2012). Comparative study of the methods used for treatment and final disposal of sewage sludge in European countries. Waste management, 32(6), 1186-1195.

Kocbek E, Garcia HA, Hooijmans CM, Mijatović I, Lah B, Brdjanovic D. (2020). Microwave treatment of municipal sewage sludge: Evaluation of the drying performance and energy demand of a pilot-scale microwave drying system. Science of The Total Environment, 742, 140541.

Kroiss H, Zessner M. (2007). Ecological and economical relevance of sludge treatment and disposal options.

Kumar C, Joardder M, Farrell TW, Karim M. (2016). Multiphase porous media model for intermittent microwave convective drying (IMCD) of food. International Journal of Thermal Sciences, 104, 304-314.

Lakshmi S, Chakkaravarthi A, Subramanian R, Singh V. (2007). Energy consumption in microwave cooking of rice and its comparison with other domestic appliances. Journal of Food Engineering, 78(2), 715-722.

Lam CM, Hsu S-C, Alvarado V, Li WM. (2020). Integrated life-cycle data envelopment analysis for techno-environmental performance evaluation on sludge-to-energy systems. Applied Energy, 266, 114867.

Léonard A, Vandevenne P, Salmon T, Marchot P, Crine M. (2004). Wastewater Sludge Convective Drying: Influence of Sludge Origin. Environmental Technology, 25(9), 1051-1057.

Lyng JG, Arimi JM, Scully M, Marra F. (2014). The influence of compositional changes in reconstituted potato flakes on thermal and dielectric properties and temperatures following microwave heating. Journal of Food Engineering, 124, 133-142.

Ma R, Yuan N, Sun S, Zhang P, Fang L, Zhang X, Zhao X. (2017). Preliminary investigation of the microwave pyrolysis mechanism of sludge based on high frequency structure simulator simulation of the electromagnetic field distribution. Bioresource technology, 234, 370-379.

Mawioo PM, Garcia HA, Hooijmans CM, Velkushanova K, Simonič M, Mijatović I, Brdjanovic D. (2017). A pilot-scale microwave technology for sludge sanitization and drying. Science of the Total Environment, 601, 1437-1448.

Mishra RR, Sharma AK. (2016). Microwave–material interaction phenomena: heating mechanisms, challenges and opportunities in material processing. Composites Part A: Applied Science and Manufacturing, 81, 78-97.

Mujumdar AS. (2014). Handbook of industrial drying (4th ed.): CRC press.

Ni H, Datta A, Torrance K. (1999). Moisture transport in intensive microwave heating of biomaterials: a multiphase porous media model. International Journal of Heat and Mass Transfer, 42(8), 1501-1512.

Ohm T-I, Chae J-S, Kim J-E, Kim H-k, Moon S-H. (2009). A study on the dewatering of industrial waste sludge by fry-drying technology. Journal of hazardous materials, 168(1), 445-450.

Peleg M. (2020). Models of Sigmoid Equilibrium Moisture Sorption Isotherms With and Without the Monolayer Hypothesis. Food Engineering Reviews, 12(1), 1-13.

Penfield MP, Campbell AM. (1990). CHAPTER 6 - INTRODUCTION TO FOOD SCIENCE. In Penfield MP, Campbell AM (Eds.), Experimental Food Science (Third Edition) (pp. 97-129). San Diego: Academic Press.

Pickles C, Gao F, Kelebek S. (2014). Microwave drying of a low-rank sub-bituminous coal. Minerals Engineering, 62, 31-42.

Pitchai K, Birla SL, Subbiah J, Jones D, Thippareddi H. (2012). Coupled electromagnetic and heat transfer model for microwave heating in domestic ovens. Journal of Food Engineering, 112(1), 100-111.

Poyet S, Charles S. (2009). Temperature dependence of the sorption isotherms of cement-based materials: Heat of sorption and Clausius–Clapeyron formula. Cement and Concrete Research, 39(11), 1060-1067.

Schulz R, Ray N, Zech S, Rupp A, Knabner P. (2019). Beyond Kozeny–Carman: Predicting the Permeability in Porous Media. Transport in Porous Media, 130(2), 487-512.

Soltysiak M, Erle U, Celuch M. (2008). *Load curve estimation for microwave ovens: experiments and electromagnetic modelling.* Paper presented at the MIKON 2008-17th International Conference on Microwaves, Radar and Wireless Communications.

Sousa KAd, Resende O, Carvalho BdS. (2016). Determination of desorption isotherms, latent heat and isosteric heat of pequi diaspore. Revista Brasileira de Engenharia Agrícola e Ambiental, 20(5), 493-498.

Souza SJFd, Alves AI, Vieira ÉNR, Vieira JAG, Ramos AM, Telis-Romero JJFS, Technology. (2015). Study of thermodynamic water properties and moisture sorption hysteresis of mango skin. 35(1), 157-166.

Stuerga D. (2006). Microwave-material interactions and dielectric properties, key ingredients for mastery of chemical microwave processes (Vol. 2): WILEY-VCH Verlag GmbH & Co. KGaA.

Vaxelaire JJJoCT, Biotechnology: International Research in Process E, Technology C. (2001). Moisture sorption characteristics of waste activated sludge. 76(4), 377-382.

Vesilind PA. (1994). The role of water in sludge dewatering. Water Environment Research, 66(1), 4-11.

Wiśniowska E, Grobelak A, Kokot P, Kacprzak M. (2019). 10 - Sludge legislation-comparison between different countries. In Prasad MNV, de Campos Favas PJ, Vithanage M, Mohan SV (Eds.), Industrial and Municipal Sludge (pp. 201-224): Butterworth-Heinemann.

5

Novel semi-decentralised mobile system for the sanitization and dehydration of septic sludge

This chapter is based on: Kocbek E, Garcia HA, Hooijmans CM, Mijatović I, Al-Addousd M, Dalala Z, Brdjanovic D. Novel semi-decentralised mobile system for the sanitization and dehydration of septic sludge: A pilot-scale evaluation in the Jordan Valley. Submitted to Journal of Environmental Science and Pollution Research, 2021.

Abstract

The provision of effective sanitation strategies has a significant impact on public health. However, the treatment of septic sludge still presents some challenges worldwide. Consequently, innovative technologies capable of an effective and efficient sludge treatment, mostly at a decentralized level, are in high demand to improve sanitation provision. To address this problem, this study evaluates a novel semi-decentralised mobile faecal sludge treatment system, the pilot-system for which consists of a combination of several individual processes including mechanical dewatering, microwave (MW) drying, and membrane filtration (ultrafiltration [UF] and reverse osmosis [RO]). The system evaluation was carried out by treating raw, partially digested faecal sludge (FS) from septic tanks – hence, septic sludge (SS) – in the Jordan Valley, Jordan. The pilot-scale system exhibited an effective and flexible treatment performance for: (i) sanitizing faecal sludge and related liquid streams; (ii) reducing the treated sludge mass (and sludge volume); and (iii) producing a high-quality treated liquid stream ideal for water reclamation applications. The mechanical dewatering process removed approximately 99% of the initial SS water content. The MW drying system completely removed E. coli and dehydrated the dewatered sludge at low energy expenditures of 0.75 MJ kg^{-1} and 5.5 MJ kg^{-1}, respectively. Such energy expenditures can be further reduced by approximately 40% by recovering energy in the condensate and burning the dried sludge, which can then be reused in land applications. The membrane filtration system (UF and RO) was able to produce high-quality treated water that is ideal for the water reuse applications that irrigation requires, as well as meeting the Jordanian standard 893/2006. In addition, the system can also be powered by renewable energy sources, such as photovoltaic energy. Therefore, this research demonstrates that the evaluated semi-decentralised mobile system is technically feasible for the in-situ treatment of SS (sanitization and dehydration), while also being effective for simultaneously recovering valuable resources, such as energy, water, and nutrients.

5.1 Introduction

Implementing effective strategies for preventing outbreaks of contagious and potentially deadly diseases is a priority for sanitation providers worldwide. Faecal sludge (FS) contain various types of pathogenic organisms such as bacteria, viruses, and parasites; therefore, they need to be properly treated and disposed in an environmentally sound manner (Jiménez et al., 2009). Their proper collection, treatment, and disposal has significantly contributed to strengthening public health (Sykes et al., 2015). Particularly, for those persons living under low-income (Ingallinella et al., 2002; Jiménez et al., 2010). Pit latrines and septic tanks are the most common alternatives worldwide for FS collection. However, such collection facilities enable only the separation of faeces from human contact providing marginal treatment; thus, if not properly treated at a later stage, such large amounts of FS only accumulate creating a subsequent waste disposal problem (Ingallinella et al., 2002; Jiménez et al., 2010; Rose et al., 2015).

FS composition is highly variable depending on the design, construction, operation and maintenance of the on-site sanitation facility where the sludge is generated and collected. Moreover, the sludge emptying frequency, temperature, rainfall patterns, and groundwater intrusion can all further influence the FS composition. For instance, FS from septic tanks, septic sludge (SS), may exhibit dry solid (DS) concentrations ranging from 0.5 to 121.6 g L^{-1}, and chemical oxygen demand (COD) from 0.4 to 91.9 g L^{-1} (Gold et al., 2018; Strande et al., 2018; Englund et al., 2020). For example, in Jordan, Halalsheh (2008) reported large fluctuations in SS composition, with DS and COD concentrations ranging from 0.8 to 10.8 g L^{-1} and from 1.4 to 14 g L^{-1}, respectively. The FS composition will determine either the final sludge disposal or the possibilities for implementing resource recovery strategies. Sludge resource recovery activities can include the reuse of the sludge in agricultural applications, and the recovery of energy via co-combustion, among others (Kacprzak et al., 2017).

Onsite sanitation facilities are either mechanically or manually desludged (Thye et al., 2011). Once collected, these large amounts of FS need to be transferred to disposal sites or treatment plants that might be located away from the points of generation; thus, increasing the overall disposal costs due to the transportation costs (Mawioo et al., 2016b). Among the available practices for a proper FS disposal is their co-treatment with municipal sewage in wastewater treatment plants (WWTPs). Such practices have a negative impact on the performance of the WWTPs, unless the WWTPs has been designed for accepting such variable loads of sludge, which is not usually the case. The sludge management costs at a WWTP may account for up to 65% of the total operational costs of the WWTPs; more than half of the sludge management costs are due to the sludge transportation costs (Wei et al., 2003; Jakobsson, 2014). These costs do not include the desludging of the septic tanks and the transporting of the FS from the onsite sanitation facilities to the WWTP.

Commonly, the FS desludging and transportation costs are covered by the owners of the on-site sanitation facilities; depending on the geographical location, such costs may represent a big fraction of the total income of a family. For instance, Murungi et al. (2014), in a study carried out in several slums in Kampala, Uganda, reported desludging and transporting costs of approximately USD 50 (for a 5 to 8 m^3 sludge track) per trip. Günther et al. (2011) reported an average monthly income of USD 36 per household in such locations. In some cases, more than one trip is required to completely empty the sanitation facility. On average, the septic tanks in Kampala, Uganda needed to be emptied once every three months up to a year (Kulabako et al., 2010). Due to such high desludging costs, the sludge is often illegally dumped; particularly, in low and middle income countries. Peal et al. (2014) reported, from a study performed in 12 cities worldwide, that only 22 % of the FS generated from on-site sanitation facilities was adequately treated and disposed, while the rest of the sludge was discarded into mostly water bodies and landfills contributing significantly to the contamination of the groundwater and endangering the public health. This situation most of the time goes unnoticed to policy makers, who usually do not take into account the informal settlements when planning the sanitation provision; that is, the adoption of a city-wide inclusive sanitation approach is commonly overlooked. As such, there is a need for a more systematic, innovative, and versatile methodologies for the proper treatment and disposal of FS ; particularly, adapting the common practices to the prevalent conditions in low and middle-income countries (Ingallinella et al., 2002).

A number of additional alternatives are available for the treatment of FS including composting, co-composting with organic solid waste, conventional drying, anaerobic digestion with or without organic solid waste, among others (Ingallinella et al., 2002; Ronteltap et al., 2014) (Mawioo et al., 2017). Some of such alternatives may convert the sludge into a valuable resource which can be further reused. However, the application of such methods is usually limited by the low throughput treatment capacity of such systems; therefore, large footprint requirements. The implementation of such processes is therefore a major challenge in areas with a pattern of rapid sludge generation and limited land availability. Those scenarios are frequently encountered in densely populated areas, such as congested cities, slums, and emergency situations. The challenge therefore, lies in finding solutions to manage FS in areas requiring responsive sludge collection, transport, and treatment (Mawioo et al., 2017).

Mobilized semi-decentralized (MSD) technologies for FS treatment have received increased attention considering the flexibility that such concept brings to the sludge treatment and management (Forbis-Stokes et al., 2021). MSD systems can be tailor-made to the needs of the users, and they can be highly suited both for short- as well as for long-term applications due to their robustness and cost-effectiveness. Furthermore, compared to centralized sludge treatment systems, MSD technologies offer additional advantages including the *in-situ* sludge treatment capacities, the ease for their deployment, and the low investment and operational costs required, among others. MSD systems must also be able to meet environmental and health standards (as well as other regulations) without exhibiting adverse effects on the environment and/or public health when treating FS. At the same time, they need to be easily scalable to address the potential rising demands driven by urbanization and population growth, while still remaining competitive compared to centralized sludge treatment methods (Afolabi et al., 2017).

In recent years, some MSD technologies have been developed for FS treatment; however, such developments have been narrowly applied in full-scale scenarios. Technologies currently available include: (i) the mobile septic tank sludge treatment system composed of centrifuge, multimedia filters, granular activated carbon filter, cartridge filters and ultrafiltration (UF)

membranes (Forbis-Stokes et al., 2021), and (ii) the LaDePa system (Septien et al., 2018). These previously developed technologies may facilitate the sludge sanitization and/or volume reduction. They have small footprint needs. Thus, they can be designed to be movable. In addition, they can be operated in non-continuous regimes. However, these technologies experienced some difficulties handling such irregular sludge flow patterns received from the on-site sanitation facilities. For instance, the mobile septic tank treatment unit as reported by Forbis-Stokes et al. (2021) has been effective on the treatment of SS with DS concentration of approximately 0.4%. On the other hand, the Ladepa sludge treatment system can only treat FS with a DS content higher than 15% due to the inherent design issues related to the performance of the perforated belt conveyor. These results have shown that the mobile treatment units currently under development have been designed for very specific conditions and their use in broader and more flexible operations is still hindered. That is, there is not a single MSD treatment system able to effectively and efficiently treat (dry and inactivate) FS at full-scale, and probably a combination of several technologies may be needed for the provision of an effective and efficient treatment.

Microwaves (MWs) are a form of electromagnetic waves with wavelengths ranging from 1 to 1,000 mm and frequencies from 0.3 to 300 GHz. MWs induce the rotation of charged ions and water molecules generating friction; thereby, converting electromagnetic energy into thermal energy. The MW technology has already been proven to be a feasible alternative to effectively dehydrate and reduce/eliminate the pathogen content of various types of sludge including blackwater sludge, SS, fresh FS, and sewer municipal sludge. The MW energy could effectively dry the sludge reaching DS concentrations of up to 90%; in addition, the inactivation of several pathogens has been demonstrated including *E. coli, Ascaris lumbricoides eggs, Staphylococcus aureus*, and *Enterococcus faecalis*, among others (Hong et al., 2004a; Hong et al., 2006; Pino-Jelcic et al., 2006; Mawioo et al., 2016a; Mawioo et al., 2016b; Mawioo et al., 2017). In MW heating, the substance being irradiated is penetrated by the MWs, so the heat generation and propagation occurs from the centre of the material outwards. Hence, making the heating process more efficient (Stuerga, 2006; Kumar et al., 2016). This substantially enhances both the efficiency, as well as the throughput capacity of the MW sludge treatment systems compared to conventional thermal dryers. Thus, paving the way for the design of more compact and MSD systems. In addition, MW systems can be immediately started-up and shut-down (without the needs of pre-heating the system before drying or cooling-down after drying); moreover, MW systems reduce the energy losses with the environment (MWs selectively heat polar materials without heating the stainless steel of the MW drying chamber, the air, or any other component of the system).

The sludge thermal sanitization and drying demands rather a lot energy. The total amount of energy required to reach the desired DS content in the sludge mainly depends on the initial amount of water present in the sludge to be evaporated - dried (Léonard et al., 2004; Mawioo et al., 2017). As such, the use of a relatively simple and low energy demanding preliminary sludge dewatering process (such as mechanical dewatering) before applying the more energy demanding processes (such as MW heating) can result in a cost-effective alternative. Such strategy has been performed and proven quite effective both at a centralized and decentralized wastewater treatment facilities (Schaum et al., 2010; Nikiema et al., 2014). Mechanical sludge dewatering processes separate the water from the sludge through the use of screens, screws, belt presses, pressure filters, centrifuges, and vacuum filters, among others; thereby, increasing the DS content at a relatively low capital and operational cost. The resulting sludge after mechanical dewatering may exhibit a DS content of up to a maximum of approximately 30%. This also lead to a considerable volume reduction of the dewatered sludge of approximately 90% (Schaum et al., 2010; Nikiema et al., 2014). The energy and operational costs savings introduced by such a reduction of both the water content as well as the volume of the sludge prior to the MW drying may counteract the initial investment and operating costs associated with such mechanical drying process.

The water extracted from the sludge both by the mechanical sludge dewatering, as well as by the MW drying processes can be later recovered for subsequent water reclamation applications. Such liquid streams contain (in addition to water) other valuable resources such as nutrients. However, they may also contain pathogenic organisms; particularly, the water rejected from the mechanical dewatering processes. Therefore, such streams must be treated before recovering those valuable resources to prevent either health risks or environmental pollution. Direct membrane UF using ceramic membranes have been evaluated for the direct treatment of raw municipal wastewater and greywater, showing a good performance in terms of the quality of the produced water (permeate) (Kramer et al., 2015; Das et al., 2018; Kramer et al., 2020). Direct membrane UF can be operated in a non-continuous regime, and it requires a relatively small footprint. However, the direct membrane UF of such streams (containing a high solids and organics concentrations, among others) exhibits some difficulties on the operation of the membranes; membrane fouling is commonly observed resulting in an increase in the capital and operational operating costs, and also reducing significantly the amount of produced clean water (permeate) (Kramer et al., 2015). The combined liquid stream produced by both the mechanical dewatering of the sludge together with the condensate from the MW drying processes contain relatively low concentrations of total solids and organics. Therefore, the fouling potential of direct ceramic membrane UF for such application may not be that significant. Moreover, several membrane processes can be accommodated in series. For instance, the permeate obtained from the direct ceramic membrane UF can be further treated by a reverse osmosis (RO) process. Therefore, obtaining a superior produced water quality both free of suspended solids, bacteria, and viruses, and also rejecting soluble compounds such as ammonium (NH_4^+), orthophosphate (PO_4^{3-}), and even ions. These compounds can be recovered in the concentrate of the RO membrane filtration for their subsequent reuse for instance as crop fertilizers. As such, the application of UF in conjunction with RO pressure-driven membranes is highly recommended for not only treating the water streams, but also for recovering resources (Kramer et al., 2015; Hube et al., 2020). This strategy is particularly relevant for regions facing water shortages, such as countries in the Middle East, as the recovered water (and other resources) represents a valuable water source for irrigation applications.

The use of specific treatment technologies for sludge treatment – such as MW treatment, mechanical dewatering systems, and membrane separation processes – has been individually demonstrated in specific applications; in particular, in the framework of centralized municipal wastewater treatment plants (Bennamoun et al. 2016, Chen et al. 2014, Das et al. 2018, Kramer et al. 2015, Kramer et al. 2020, Mawioo et al. 2016b). However, the combination of such treatment technologies to aim for the development of an improved decentralized FS treatment alternative has not been yet carried out. In addition, the validation of such a combination of technologies in real-life scenarios is still pending (Forbis-Stokes et al. 2021, Nikiema &Cofie 2014, Septien et al. 2018). Furthermore, the proposed concept (train of technologies) is envisioned to be very energy demanding (particularly the MW drying units) and this could be challenging in low-income countries and/or in humanitarian settings. However, there are alternatives beyond powering the system by connecting to the local grid. Such alternatives include the provision of diesel generators, as well as powering the system using renewable sources such as solar/PV energy.

The present study is aimed at investigating the feasibility of a novel pilot-scale MSD unit for the *in-situ* treatment of SS (drying, pathogen inactivation and liquid streams treatment). The novel pilot-scale MSD unit included an mechanical dewatering unit, an MW drying system, and a membrane filtration system (UF and RO), while the feasibility of the evaluated technologies for resources recovery was also explored. The sludge treatment performance was assessed, first by characterizing the treated sludge and generated liquid streams properties, and, second, by determining the treatment performance of each individual set of technologies conforming to the pilot-scale system. Following this, the validation of the proposed technology was carried out in the Jordan Valley area by treating locally generated SS.

5.2 Material and methods

5.2.1 Experimental unit

The MSD pilot-scale unit was designed and manufactured by Tehnobiro d.o.o (Maribor, Slovenia). The system performance was evaluated at the experimental facilities of the German Jordanian University (GJU) in the Jordan Valley area, Jordan (Figure 36), with a focus on treating SS. The pilot system consisted of a combination of three key components: an mechanical dewatering unit, an MW drying system, and a membrane filtration (UF and RO) system. The system components were designed to be operated in a batch mode, allowing for robust process control and flexibility. The pilot system was built in a standard (six-meter length) cargo-size container and mounted on a trailer to facilitate the unit's mobility. The pilot system was designed for the treatment of approximately 130 L h^{-1} of SS from an initial DS concentration of approximately 2% to a final DS concentration of 85%. The evaluation of the pilot system was carried out in the winter season and the key prototype components are displayed in Figure 37. A detailed description of all the system components is presented in Sections 5.2.1.1 to 5.2.1.6.

Figure 36: Project location at the Jordan Valley - geographical coordinates 31° 54'38.78"N and 35° 34'40.63"E (source: Google Maps)

(a)

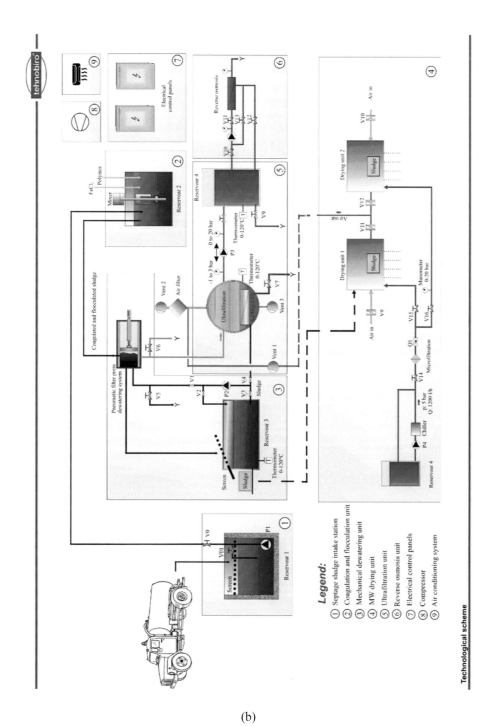

(b)

Figure 37: (a) General view of the pilot system; and (b) schematic representation (Tehnobiro d.o.o)

5.2.1.1 Septic tank sludge intake station

The SS intake station, show in Figure 38 (#1 in Figure 37) consisted of a 500 L polyethylene receiving tank, a rough screen, and a submersible pump with a floating level switch.

Figure 38: Septic tank sludge intake station

5.2.1.2 Coagulation and flocculation unit

The coagulation and flocculation unit (#2 in Figure 37) consisted of a reservoir and a mixer. The unit was placed right after the intake station and before the mechanical dewatering unit. The coagulation agent (ferric chloride) and flocculation agent (polymer (Drewlock™ 469) based on acrylamide and a cationic co-monomer (Solenis, USA) were manually added into the unit to promote particle destabilization and flocculation of the suspended solids in the SS. In such a way, larger particles were formed that could be better separated in the downstream mechanical dewatering unit. The coagulation and flocculation method is described in Section 5.2.2.

5.2.1.3 Mechanical dewatering unit

The mechanical dewatering unit shown Figure 39 (#3 Figure 37) consisted of a 500 L equalization tank (500x1000x1000 mm) equipped with a rough screen (30 mm), a water level transmitter, a pneumatic filter press dewatering system, and a centrifugal pump with an open impeller. The pneumatic filter press dewatering system consisted of a cylindrical stainless-steel chamber with a diameter of 125 mm and length of 300 mm. Inside that cylindrical chamber a 0.5 mm slotted cylindrical screen was located. A stainless-steel piston pneumatically operated pushed the sludge through the cylindrical screen and in such a way dewatered the sludge. Compressed air was supplied to the pneumatic unit by a compressor (#8 Figure 37). The sludge dewatering unit produced two outputs: (i) the concentrated solid sludge cake, and (ii) the liquid filtrate. The concentrated sludge cake was further treated (dried) at the MW drying unit (Section 5.2.1.4). The filtrate was further treated by the membrane separation system (Section 5.2.1.5). As indicated in Figure 37, the mechanical dewatering unit was also provided with a centrifugal pump for bringing the concentrate from the downstream membrane filtration system back to the mechanical dewatering unit. The higher the DS concentration in the UF concentrate, the higher the opportunities to treat the UF concentrate directly in the MW drying unit. The fouling properties of the UF concentrate determined the maximum achievable DS concentrations in the UF concentrate, while the UF membrane fouling was monitored by measuring the pressure drop across the membrane.

Figure 39: Mechanical dewatering unit

5.2.1.4 MW thermal unit

The MW drying unit, shown in Figure 40 (#4 Figure 37) was designed to sanitize and dry the mechanically dewatered sludge by using MW radiation. The drying unit consisted of two cylindrical MW cavities, rotating polypropylene (PP) turntables, sludge holding vessels (PP), a ventilation unit for vapour extraction, an air filtration system, two MW power supply sources, and two MW magnetrons with a power output capacity of up to 6 kW each, operated at a frequency of 2,450 GHz. The cavities were made of stainless steel with a diameter of 400 mm and a total length of 400 mm. The PP turntables were used to slowly and continuously rotate the irradiated sludge at a speed of one rpm; thus, alleviating the effect of potential non-uniform temperature distributions. The PP holding vessels were designed for a maximum load of 6 kg each. The exhaust vapour extracted by the ventilation system was treated by the air filtration system composed of: (i) activated carbon soaked in phosphoric acid, (ii) activated carbon soaked in sodium hydroxide, and (iii) aluminium oxide with potassium permanganate. The MW magnetrons and power supply sources were cooled down by recirculating demineralised water at a flowrate of 1,200 L h^{-1} at a pressure of 5 bar. The changes in the moisture content of the sludge during the MW drying were continuously measured by a single point load cell (Mettler Toledo). The temperature of the sludge was measured intermittently during the sanitization process using a thermal camera (FLIR TG 165, FLIR Systems Inc., USA) and a temperature gauge (ELPRO Lepenik & Co., Slovenia). The detailed description of the MW unit is found in chapter 2.

Figure 40: Drying unit

5.2.1.5 Membrane separation system

The membrane filtration treatment system is presented in Figure 41 (#5 and #6 in Figure 37). The sludge filtrate from the mechanical dewatering, together with the condensate produced from the evaporated water from the MW drying unit, were collected in the UF reservoir prior to the UF process. Within this reservoir, a ceramic UF membrane element with an average pore size of 0.08 μm and a total filtration area of 1.05 m^2 was immersed and operated in a dead-end filtration mode for permeate extraction. An air diffuser manifold was installed at the bottom of the UF reservoir, just below the ceramic membranes, for providing membrane scouring. The UF system was equipped with a peristaltic pump to extract the treated permeate out of the system and to backwash the membranes. The pump was provided with a variable frequency drive, so different flowrates could be established for either a permeate production at 24 L h^{-1}, or a backwash flowrate at 47 L h^{-1} (i.e., setting operational fluxes of 23 and 45 L m^{-2} h^{-1} for the permeate production and backwash, respectively). The membrane backwash was applied for 60 seconds every 10 minutes. The operating temperature and pressure were monitored by a temperature and a pressure gauge (ELPRO Lepenik & Co., Slovenia). The permeate obtained from the UF membranes was stored in a permeate tank from which it was subsequently pumped to an RO membrane filtration unit for further treatment. A booster pump was used to drive the UF permeate through a single RO membrane element unit (Lenntech type CSM RE 4021-BE) with a nominal salt rejection capacity of 99.7%. The RO membrane unit had an effective membrane area of 3.3 m^2, and it was operated in a crossflow mode at a flowrate of 230 L h^{-1} (i.e., a flux of 70 L m^{-2} h^{-1}) and at a pressure of 8 bars. The RO membrane element was inserted into a compact pressure housing vessel. The pressure was monitored in the feed line by installing a pressure gauge (ELPRO Lepenik & Co., Slovenia). An additional pressure gauge was added in the concentrate line to monitor the changes in the transmembrane pressure.

The RO permeate recovery was adjusted to 56% by means of a butterfly valve located at the membrane's feed line. Three sampling points were provided as follows: in the feed, in the concentrate, and in the permeate lines.

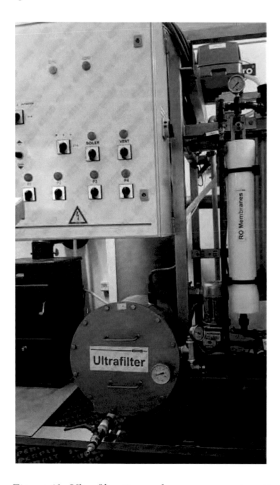

Figure 41: Ultrafiltration and reverse osmosis unit

5.2.1.6 Other components

Other components included: (i) the electrical control panels provided for the operation of the system (#7, Figure 37), (ii) a compressor (#8, Figure 37), and (iii) an air conditioning system for ambient temperature control (#9, Figure 37).

5.2.2 Experimental procedure

The SS was obtained from a lined septic tank (3x3x4 m) located in the proximity of the GJU experimental site in the Jordan Valley area of Jordan. The septic tank collected and partially treated waste that was primarily generated by a nearby hospital, with the SS emptying frequency varying between two and three months. A groundwater well was located approximately 50 m away from the septic tank and, on three different occasions (three different batches to suit this research), the SS was transported to the experimental site by an SS pump-out truck. The SS was then conveyed to the receiving reservoir (#1 in Figure 37), before subsequently being pumped to the coagulation and flocculation unit (#2 in Figure 37). Within this unit, the suspended solids were destabilized and flocculated by adding ferric chloride and powder polymer. The required dosages of ferric chloride and polymer were determined onsite by carrying out jar-test experiments for each individual SS batch. A ferric chloride solution was added to a one L beaker containing the SS. The solution was stirred for five minutes at 100 rpm. After destabilizing the solids, the polymer was introduced while gently stirring the solution for 10 minutes at 50 rpm. The SS flocs were then allowed to settle for ten minutes. After settling, the jar's contents were passed through a 0.5 mm sieve (same diameter as the sieve contained in the mechanical dewatering unit described in Section 5.2.1.3.). The ferric chloride and polymer optimal doses were determined by evaluating the sieve sludge retention efficiency and weighing the mass of the dewatered SS retained on the filter. In addition, visual inspections were carried out to assess both the consistency of the SS flocs, and, the remaining turbidity of the supernatant. SS flocs with a slimy appearance and consistency indicated a polymer overdose. The polymer optimal dose determination was carried out with the assistance of the polymer manufacturer (Specialized Water Technologies, Amman, Jordan).

Consequently, the flocculated SS was conveyed to the mechanical dewatering unit (#3, Figure 37) by gravity. Two products were obtained out of the mechanical dewatering process: (i) the dewatered SS and (ii) the filtrate. The dewatered SS was manually collected in PP vessels and transferred to the MW drying unit (#4, Figure 37), whilst the liquid stream (filtrate) was further treated in the membrane filtration system (#5 and #6, Figure 37). The dewatered SS (SS cake) obtained from the mechanical dewatering unit was manually collected and weighted. Then, the dewatered SS was placed in the MW drying unit holding vessels. The dewatered SS was irradiated in the MW units in batches of 0.5 kg each, with the power set at 1.5 kW. The MW drying units were operated long enough (for approximately 26 minutes) to provide sufficient energy for both inactivating the pathogens and dehydrating the SS. The MW irradiation caused the SS temperature to increase, which led to the subsequent sanitization of the sludge and evaporation of the water. The evaporated water (vapour) was directed to a filtration system for odour control by means of a ventilator. The resulting condensate was collected in the UF reservoir together with the filtrate produced by the mechanical dewatering unit. These liquid streams, both the condensate and the filtrate, were filtered by the UF ceramic unit. The resulting permeate was further treated (polished) by the RO unit.

Several different samples were collected to represent the entire treatment process as numbered in Figure 43: (1) SS; (2) Filtrate; (3) Mechanically dewatered sludge; (4a) Sanitized sludge; (4b) Sanitized and dried sludge; (5) Condensate; (6) UF permeate; (7) UF concentrate; (8) RO permeate; and (9) RO concentrate. Three independent evaluations (batches) were carried out, with one individual sample taken from each batch at the sampling points indicated in Figure 43. The analytical determinations conducted on the collected samples are presented in the following Section 5.2.3.

Figure 42: Septic tank sludge conveyance (photo taken by H.A. Garcia Hernandez)

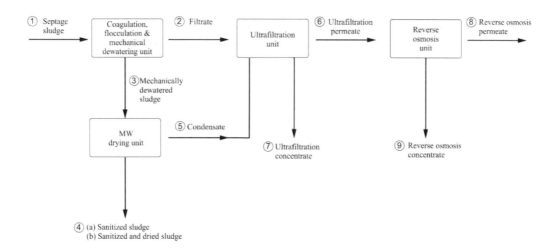

Figure 43: Process flow diagram indicating the sampling points as follows: (1) Septic tank sludge; (2) Filtrate; (3) Mechanically dewatered sludge; (4.a) Sanitized sludge; (4.b) Sanitized and dry sludge; (5) Condensate; (6) UF permeate; (7) UF concentrate; (8) RO permeate and; (9) RO concentrate.

5.2.3 Analytical determinations

The samples, as described in Figure 43, were collected in polyethylene bottles tightly sealed with a cap, then stored in a refrigerator. The reservoirs containing sludge at low DS concentrations (such as the raw SS reservoir) were well mixed before sampling with either an air or a recirculation pump, carried out to reduce the possibility of solids settling and, thus, allowing to obtain a representative sample. The reservoirs containing sludge at high DS concentrations (such as the dewatered SS reservoir) were mixed manually before sampling. The sludge samples were taken approximately one hour after each particular individual treatment process was initiated, then were properly preserved to allow for the specific physical-chemical parameters to be determined. The physical-chemical characteristics of the collected samples were analysed to determine the levels of DS, volatile solids (VS), *E. coli*, chemical oxygen demand (COD), soluble COD (sCOD), total nitrogen (TN), total phosphorus (TP), PO_4^{3-}, NH_4^+, potassium (K), heavy metals (chromium (Cr), nickel (Ni), cadmium (Cd), Zinc (Zn), Lead (Pb), and Mercury (Hg), and elemental analysis including the following elements: carbon (C), hydrogen (H), nitrogen (N) and oxygen (O). The analytical determinations were carried out at an external laboratory in Jordan (NaraTech LABs, Amman, Jordan), with the samples analysed within two days of sample collection. Approximately, nine samples were collected for each batch (i.e., 27 samples in total). Table 7 describes the different analytical determinations carried out on the different samples, while the methods conducted for the analytical determinations are shown in Table 8. The pH and the electrical conductivity (EC) of the samples were analysed immediately after the samples were taken using a portable pH (ProfiLine pH 3110, WTW, Germany) and an EC meter (GMH 3430, Greisinger, Germany). The final DS concentration of the dried sludge was calculated by determining the mass change of the samples (as they were being dried in the MW unit). The mass was measured using a single point-cell (scale) installed within the MW cavities. The higher heating value (HHV), or gross energy, was calculated from the elemental composition of the sludge as per Friedl et al. (2005). All the results were reported in consideration of the average values and standard deviations from the three different batches. Each SS batch required two days to be fully processed, thus, each batch corresponded to two days of system operation. The following physical-chemical parameters (E. coli, TP, and PO_4^{3-}) were calculated by conducting mass balances (i.e., those parameters were not analysed).

Table 7: Monitoring parameters

Sample point and ID		DS	VS	E.coli	COD	sCOD	TN	NH$_4^+$	TP	PO$_4^{3-}$	K$^+$	pH	Conductivity	Heavy metals	CHNO
1.	Septic tank sludge	•	•	•	•	•	•	•	•	•	•	•	•		
2.	Sludge filtrate	•	•		•	•	•	•	•	•	•	•	•		
3.	Mechanically dewatered sludge	•	•	•	•	•	•	•		•				•	•
4.a	Sanitized sludge			•											
4.b	Sanitized and dried sludge	•		•											
5.	Condensate				•	•	•	•	•	•	•	•	•		
6.	UF permeate	•		•	•		•	•	•		•		•		
7.	UF concentrate	•			•	•	•	•	•	•	•	•	•		
8.	RO permeate	•			•	•	•	•	•	•	•	•	•		
9.	RO concentrate	•			•	•	•	•	•	•	•	•	•		

Table 8: Analytical methods for the determination of the sludge physical-chemical characteristics

Parameter	Method
E. coli	ISO9308:2014
COD	SM5220C
sCOD	SM5220C
TN	EL-WL-SOP-003
TP	ST066
PO$_4^{3-}$	SM4110
NH$_4^+$	SM4110
Metals	ST066
TS	SM-2540D
VS	SM-2540E
CHNO	ST078

5.3 Results and discussion

5.3.1 Performance of the coagulation/flocculation and mechanical dewatering units on the treatment of SS

The characteristics of the SS before being flocculated (influent SS) and after mechanical dewatering are presented in Table 9. The particle destabilization of the sludge was carried out by dosing ferric chloride at 120 mg L^{-1} followed by rapid mixing, while the flocculation of the sludge was achieved by adding the polymer at a concentration of 50 mg L^{-1} followed by slow mixing. The pilot unit was operated in a batch mode; an overall mass balance of the coagulation/flocculation and mechanical dewatering unit is presented in Figure 44.

The DS concentration in the influent SS, as shown in Table 9, was 0.16 % on average. This value agrees with the literature on sludge derived from septic tanks in Jordan (Halalsheh 2008). For example, Halalsheh (2008) reported average DS content values of 0.2% in SS samples examined in the winter and 0.5% in those examined in the summer. The low DS values reported in the SS could be partly attributed to a pilot test period (winter season) characterized by higher precipitation rates than during the summer season. Brackish groundwater intrusion into septic tanks may also contribute to such low DS values depending on the rain precipitation patterns, water table levels, and construction characteristics of the septic tanks (Halalsheh 2008). The EC measured in the influent SS (reported in Table 9) exhibited a value of 2.9 x 10^3 μS cm^{-1}. As estimated by applying the software Hydranautics IMS, the presence of NH_4^+, PO_4^{3-} and K^+ contributed to a conductivity of 1.0 x 10^3 μS cm^{-1} (approximately 34% of the influent SS conductivity) equivalent to a total dissolved solids (TDS) concentration of 0.03%. Moreover, according to the formula provided by Rusydi (2018) (TDS = k x EC considering a correlation coefficient k of 0.55), the TDS values of the remaining ions were approximately 0.1%. Thus, these results indicated that a major component of the total DS measured in the SS (0.16%) were dissolved solids (0.13%), meaning that the SS contained a relatively low amount of suspended solids (0.03%). Based on these previous calculations, an eventual brackish groundwater intrusion in the influent SS can be suspected. The intrusion of brackish groundwater in SS tanks in Jordan was confirmed by Halalsheh (2008); the author reported an average TDS concentration of approximately 0.1 and 0.3% in SS samples analysed in the winter and summer period, respectively. Such TDS content in the SS is close to the maximum allowable limit for water reuse for irrigation purposes (TDS <0.15% or <1.5 g L^{-1}) as stated in the Jordanian standard 893/2006 (Ulimat 2012). Accordingly, if the water content of such SS is intended to be applied in irrigation applications, then it would be safe to apply further measures, such as RO filtration for removing divalent and monovalent ions. The type of wastewater the septic tank is receiving (hospital wastewater and/or laundry wastewater), as well as the strategies followed for emptying the septic tank (removing just the liquid fraction – supernatant – from the septic tanks rather than settled solids) could contribute to the DS content of the SS.

The influent SS was destabilized and flocculated using ferric chloride and polymers, respectively. Upon flocculation, the SS was dewatered using the mechanical dewatering unit. The DS content of the SS went up from 0.16% (influent SS) to 5.6% (dewatered SS). The influent SS was concentrated approximately 35 times indicating the fairly good performance of the mechanical dewatering unit. However, the DS concentration obtained in the dewatered SS were much lower than expected considering both the initial low DS content of the influent SS, as well as the high presence of TDS that ended up in the filtrate. As shown in the mass balance presented in Figure 44, approximately 13% of the DS initially present in the influent SS were retained by the filter screen of the mechanical dewatering unit, while the rest of the DS content pass through the sludge filtrate. The relatively high content of DS in the sludge filtrate, as previously reported, could be due to the high presence of TDS (not measured in this research but estimated from the EC determinations). In addition, the passage of such a large amount of solids into the filtrate could also be due to the relatively large aperture size of the sieve in the mechanical dewatering unit (of approximately 0.5 mm). Due to the sieve having such a large aperture size, negligible clogging of the sieve was observed.

The coagulation/flocculation process could promote the simultaneous precipitation and removal of nutrients (such as phosphate) and organics sequestered in the solid cake. For instance, as shown in Figure 44 the total mass of TP and PO_4^{3-} was much higher in the dewatered sludge (3.5 g of TP – 95% of the influent content, and 2.5 g of PO_4^{3-} – 95% of the influent content) compared to the sludge filtrate (0.2 g of TP – 4% of the influent content and 0.1 g of PO_4^{3-} – 5% of the influent content). On the other hand, as shown in Figure 44, 83, 99, and 94% of the influent content for the sCOD, NH_4^+ and K^+, respectively remained in the liquid fraction. The results presented in Table 9 and Figure 44 indicated that the soluble forms of both nutrients (except PO_4^{3-}) and organic compounds present in the influent SS ended up in the filtrate (liquid fraction), while the particulate material were entrapped within the solid matrix as expected.

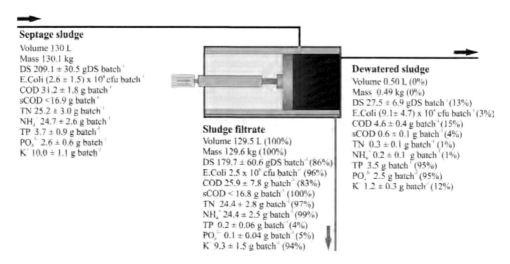

Septage sludge
Volume 130 L
Mass 130.1 kg
DS 209.1 ± 30.5 gDS batch⁻¹
E.Coli (2.6 ± 1.5) x 10⁶ cfu batch⁻¹
COD 31.2 ± 1.8 g batch⁻¹
sCOD <16.9 g batch⁻¹
TN 25.2 ± 3.0 g batch⁻¹
NH₄ 24.7 ± 2.6 g batch⁻¹
TP 3.7 ± 0.9 g batch⁻¹
PO₄ 2.6 ± 0.6 g batch⁻¹
K 10.0 ± 1.1 g batch⁻¹

Sludge filtrate
Volume 129.5 L (100%)
Mass 129.6 kg (100%)
DS 179.7 ± 60.6 gDS batch⁻¹ (86%)
E.Coli 2.5 x 10⁶ cfu batch⁻¹ (96%)
COD 25.9 ± 7.8 g batch⁻¹ (83%)
sCOD < 16.8 g batch⁻¹ (100%)
TN 24.4 ± 2.8 g batch⁻¹ (97%)
NH₄ 24.4 ± 2.5 g batch⁻¹ (99%)
TP 0.2 ± 0.06 g batch⁻¹ (4%)
PO₄ 0.1 ± 0.04 g batch⁻¹ (5%)
K 9.3 ± 1.5 g batch⁻¹ (94%)

Dewatered sludge
Volume 0.50 L (0%)
Mass 0.49 kg (0%)
DS 27.5 ± 6.9 gDS batch⁻¹ (13%)
E.Coli (9.1± 4.7) x 10⁴ cfu batch⁻¹ (3%)
COD 4.6 ± 0.4 g batch⁻¹ (15%)
sCOD 0.6 ± 0.1 g batch⁻¹ (4%)
TN 0.3 ± 0.1 g batch⁻¹ (1%)
NH₄ 0.2 ± 0.1 g batch⁻¹ (1%)
TP 3.5 g batch⁻¹ (95%)
PO₄ 2.5 g batch⁻¹ (95%)
K 1.2 ± 0.3 g batch⁻¹ (12%)

Figure 44: Mass balance of the coagulation/flocculation unit and mechanical dewatering

Table 9: SS physical-chemical characteristics before and after coagulation/flocculation and mechanical dewatering

Parameter	Units	Septic tank sludge (influent)	Filtrate	Mechanically dewatered sludge
DS	%	0.16 ± 0.02	0.14 ± 0.05	5.62 ± 0.14
VS	%	47.1 ± 9.60	50.7 ± 1.18	31.8 ± 0.67
Density[b]	g cm^{-3}	1.00	1.00	1.02
pH	-	7.5 ± 0.21	7.3 ± 0.42	6.5 ± 0.22
EC	µS cm^{-1}	$(2.94 \pm 0.24) \times 10^3$	$(3.05 \pm 0.21) \times 10^3$	-
E. coli[c]	CFU g^{-1}	$(2.00 \pm 1.15) \times 10^1$	1.93×10^{1a}	$(1.87 \pm 0.97) \times 10^2$
COD[c]	mg g^{-1}	0.24 ± 0.01	0.20 ± 0.06	9.48 ± 0.75
sCOD[c]	mg g^{-1}	<0.13	<0.13	-
TN[c]	mg g^{-1}	0.19 ± 0.02	0.19 ± 0.02	0.56 ± 0.27
NH$_4$$^{+}$[c]	mg g^{-1}	0.19 ± 0.02	0.19 ± 0.02	0.40 ± 0.26
TP[c]	mg g^{-1}	0.03 ± 0.01	0.001 ± 0.0005	7.25^a
PO$_4$$^{3-}$[c]	mg g^{-1}	0.02 ± 0.005	0.001 ± 0.0003	5.10^a
K^{+}[c]	mg g^{-1}	0.08 ± 0.01	0.07 ± 0.01	2.40 ± 0.51
E. coli[d]	CFU gDS^{-1}	$(1.2 \pm 0.7) \times 10^4$	1.4×10^{4a}	$(3.3 \pm 1.7) \times 10^3$
COD[d]	mg gDS^{-1}	149.2 ± 8.8	144.0 ± 43.2	168.7 ± 13.4
sCOD[d]	mg gDS^{-1}	<80.8	<93.6	-
TN[d]	mg gDS^{-1}	120.6 ± 14.5	135.8 ± 15.5	10.0 ± 4.8
NH$_4$$^{+}$[d]	mg gDS^{-1}	118.1 ± 12.4	135.6 ± 14.04	7.2 ± 4.6
TP[d]	mg gDS^{-1}	17.8 ± 4.1	0.9 ± 0.3	129.0^a
PO$_4$$^{3-}$[d]	mg gDS^{-1}	12.6 ± 3.1	0.7 ± 0.2	90.9^a
K^{+}[d]	mg gDS^{-1}	47.7 ± 5.2	51.8 ± 8.2	42.6 ± 9.1

[a] calculated from the mass balance
[b] the density of sludge obtained from Radford et al. (2014)
[c] the concentrations of chemical compounds or colony forming units were expressed as mg or CFU per g of solution
[d] the concentrations of chemical compounds or colony forming units were expressed as mg or CFU per g of DS (dry mass basis)

The mechanical dewatering process allowed the removal of approximately 99.6% of the initial SS water content. Thus, the subsequent MW drying unit would receive: (i) less volume of sludge requiring drying; and (ii) sludge with a much lower water content (i.e., much higher initial DS%) compared to the influent SS. In this study, each batch of sludge to be MW dried had a volume of approximately 0.5 L at a DS% of 5.6% compared to the SS influent batch before dewatering with an initial total volume of 130 L at a DS of 0.16%. As such, the MW power output needed to meet the treatment objectives (such as the final DS% of the dried sludge) can be optimized. This will have a positive impact on the overall capital costs, operational costs, and footprint needs of the system. These results demonstrated that pre-treating the sludge with the mechanical dewatering process had a positive effect on the subsequent MW drying process. In addition, a liquid stream is produced (filtrate) that can be further polished obtaining different water qualities for further water reclamation applications.

5.3.2 Performance of the MW drying unit on sludge sanitization, drying, and suitability of the sanitized sludge for land applications

This section presents and evaluates the effect of the MW radiation on the sanitization and drying performance of the dewatered SS retained by the mechanical dewatering unit. The sanitization and drying performance of the MW system were assessed by monitoring the pathogenic content of the sludge and the changes in sludge mass, respectively, as a function of the exposure time at predefined MW power outputs. The final products resulting from the MW treatment including the treated (sanitized and dried) sludge and the water evaporated (condensate) during the MW heating were characterized for evaluating the presence/absence of contaminants that may eventually have a negative effect on crops. In addition, other components in the dried sludge were characterized such as the calorific value and the elemental composition of selected elements.

5.3.2.1 Sanitization of sludge using MW-based technology

Figure 45 illustrates the MW performance for the sanitization of the mechanically dewatered sludge evaluated by determining the *E. coli* log removal as a function of the exposure time and MW specific energy output (i.e., MW power output per unit of initial sludge mass). Figure 45 also indicates the change in the sludge temperature as a function of the exposure time. The *E. coli* log reduction increased linearly as a function of the exposure time, MW specific energy output, and temperature. The exposure time required to achieve 4.4 *E. coli* log removal was 250 seconds. That is, 250 seconds were required to reduce the *E. coli* count from 1.87×10^2 cfu g^{-1} (or 3.3×10^3 cfu gDS^{-1}) to below the limit of detection. This corresponded to a MW specific energy output of 0.75 MJ kg^{-1} (0.2 kWh kg^{-1}) at a final temperature of 80 °C. According to the results presented in Figure 45 the removal of pathogens (expressed as *E. coli* in this study) was obtained within a short exposure time and energy expenditures. This observation agrees with the literature. Several studies have been conducted on a wide range of sludge, including blackwater sludge, waste activated sludge, centrifuged waste activated sludge, SS and fresh FS (Hong et al., 2004a; Hong et al., 2006; Pino-Jelcic et al., 2006; Mawioo et al., 2016a; Mawioo et al., 2016b; 2017). The removal of pathogens could be attributed to the molecular excitation caused by MW irradiation of the sludge. This led to an increase in the temperature of the sludge, a critical factor influencing the microbial destruction in MW applications. A complete microbial destruction occurs at temperatures above 70 °C (Hong et al., 2004a). Thus, the complete inactivation of *E. coli* achieved at a temperature of 80 °C agrees with previously reported microbial removal temperatures. In addition, the excitation of the molecules caused by the MW irradiation of the sludge could have damaged the cells through non-thermal effects (Banik et al., 2003; Hong et al., 2004b). The non-thermal effects, unlike the thermal effects, cannot be directly measured. So, such effects are suspected to result from the interaction between the polarized macromolecules and the oscillating electric field generated within the MW cavity. This interaction has been reported to wriggle the polarized molecules, causing cell breakages leading to cell death and the consequent release of intracellular material (Herrero, et al., 2008). In both cases, thermal and non-thermal microbial destruction, it has been observed that the degree of cell damage increases with increasing the specific energy output. The energy required to achieve a complete pathogen destruction has been reported to vary between 0.33 MJ kg^{-1} to

3.1 MJ kg^{-1} (0.09 to 0.9 kWh kg^{-1}) (Karlsson et al., 2019). In this study, a MW specific energy output of 0.75 MJ kg^{-1} (0.2 kWh kg^{-1}) was required in good agreement with the literature. The differences in the specific energy outputs required to achieve a complete pathogen destruction may be attributed to several factors including the selected pathogen indicators, physical-chemical properties of the sludge (e.g., mass, density, specific heat, ionic content, dielectric properties), operational parameters, and specific MW related settings such as the MW operating frequency, MW power outputs, and overall MW system design features, among others. For instance, Hong et al. (2016), when microwaving coal, reported that the MW power output was one of the most critical factors to consider when designing a MW drying system. The power output had a strong influence on how fast the material was heated up. Thus, the temperature required for the sanitization of the sludge can be eventually reached more quickly by applying a higher MW power output; in addition, this would lead to lower specific energy outputs for achieving the desired goal. The applied MW power output for achieving complete sanitization of the sludge would also depend on the initial DS content of the sludge. That is, the lower the net amount of water present in the sludge, and thus the lower the specific heat capacity, the less energy would be required to reach the sanitization temperature with the corresponding pathogen destruction (Mawioo et al., 2017). The pathogen destruction may also vary according to the specific indicators used to evaluate the sanitization performance of the MW system. For instance, Mawioo et al. (2017) reported that the MW specific energy output required to inactivate *E.Coli* in SS was much larger compared to the energy required to inactivate both *enterococcus faecalis, and Staphylococcus aureus.* In addition, the type of sludge being irradiated also exhibited different pathogen removal efficiencies (Mawioo et al., 2017). In addition, the specific energy output required to sanitize the material would also depend on the MW system design (Vadivambal et al., 2010). Whereas, the most important design issue associated with pathogen destruction is the uniformity of the electromagnetic energy distribution within the material. An uneven temperature distribution would lead to an incomplete pathogen destruction; therefore, raising issues for the safe handling and potential reuse of the dried sludge (Vadivambal et al., 2010). The occurrence of hot and cold spots can be reduced by varying the MW frequency and/or by the provision of a turntable and/or an agitator in the MW irradiation cavity (Jones et al., 2002) as done in this study.

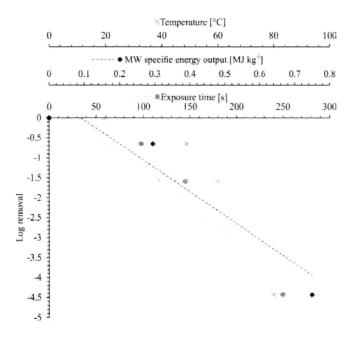

Figure 45: Log removal of E. coli as a function of exposure time, MW specific energy output, and temperature

The results presented in this study confirmed that MW irradiation can be applied for the sanitization of the sludge. The MW system was capable of achieving a complete reduction of *E. coli* at the evaluated conditions at a relatively low MW specific energy output (and short exposure times).

5.3.2.2 Drying of sludge using MW-based technology

Figure 46a illustrates the changes in the sludge moisture content (dry solids content) as a function of the MW irradiation exposure time at a MW power output of 1.5 kW. The MW irradiation chamber was loaded with 0.5 kg of sludge at a DS concentration of 5.6%, so an initial power/mass ratio of 3 kW/kg was applied. This value corresponds to the maximum initial power/mass ratio applied in previous MW pilot-scale evaluations (Mawioo et al., 2017; Kocbek et al., 2020). The sludge was irradiated for approximately 26 minutes, and the DS content increased from 5.6% to 30.0% (i.e., sludge moisture content decreased from 16.8 to 2.4 kg of water kg of dry solids^{-1}). As a result of such dehydration of the sludge, the sludge volume decreased by approximately 83% as shown in Figure 46b. This corresponds to an energy consumption 4.6 MJ kg^{-1} (1.3 kWh kg^{-1}). Notably, a higher degree of volume reduction could still be achieved by simply prolonging the irradiation exposure time. However, a substantial increase in the energy expenditures would occur. Such extrapolation is indicated in Figure 47 (provided that the energy expenditure of the system is proportional to the exposure time). So, a MW specific energy output (defined in Section **Error! Reference source not found.**) of 5.5 MJ kg^{-1} (1.5 kWh kg^{-1}) would be needed to achieve a volume reduction of approximately 95%

corresponding also to a DS content higher than 95%. Such specific energy output calculation does not consider the efficiency for converting the electrical energy into electromagnetic energy (i.e., the MW generator energy efficiency). The MW generator efficiencies utilising magnetrons range between 50 to 72% and between 80 to 90% at MW frequencies of 2,450 MHz and 915 MHz, respectively (Evans et al., 1996). As such, the energy expenditures (cost) for drying the sludge would directly depend on the type of energy (MW frequencies) used by the system. According to Kocbek et al. (2020), MW systems operated at a power output of 1.5 kW have a MW generation efficiency of 57%. Considering this value, the specific energy consumed during the drying process would be almost double to that reported in this study (8.6 MJ kg^{-1} or 2.4 kWh kg^{-1}). Moreover, as indicated in the previous Section **Error! Reference source not found.**, and also as reported by Kocbek et al. (2020), the higher the MW power output, the higher the power absorbed by the material being dried, minimizing the possibility of having energy losses somewhere in the system; therefore, increasing the energy efficiency of the MW drying system. Thus, operating the system at the highest nominal power capacity maximizes the MW energy efficiency leading to a reduction in the specific energy output required to dry the material. An increase in the rate at which the energy is delivered has a strong influence on the MW generation efficiency. In addition, the higher the energy absorbed by the material per unit of volume, the higher the water evaporation rate; therefore, the shorter required irradiation exposure times (Chen et al., 2014; Bennamoun et al., 2016). In other words, the energy expenditure of MW technology is partially dependent on the MW operational conditions, and thus it may be optimized by varying the sludge mass or the MW power output settings.

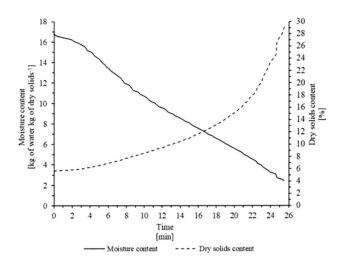

(a)

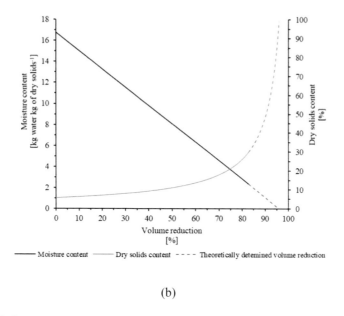

(b)

Figure 46: Sludge moisture content and dry solids content as a function of (a) exposure time and (b) sludge volume reduction

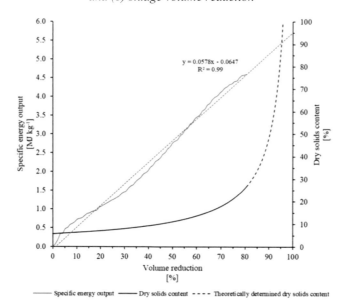

Figure 47: Specific energy output and sludge DS content as a function of the sludge volume reduction

Furthermore, the energy contained in the vapours/condensate (produced when evaporating the water content of the sludge) may be eventually used to increase the temperature of the dewatered sludge before the sludge is introduced into the MW irradiation cavity. Thus, reducing the overall energy demand for drying the sludge; the specific energy output can be reduced by approximately 10% (Kocbek et al., 2020). In addition, energy can also be recovered/obtained from burning the MW dried sludge by a combustion process. The elemental composition of the dried sludge for C, H, N and O were determined and are presented in Table 10. The gross calorific value of the dried sludge was determined from the elemental composition determination, and it is also presented in Table 10. If each batch is MW dehydrated from an initial sludge mass of 500 g (at a 5.6% DS) to a final dried sludge mass of 30 g (at 90% DS), an additional energy recovery of 0.5 MJ (i.e., 30 grams at 16 MJ kg^{-1}) or specific energy output savings of 1 MJ kg^{-1} can be achieved by burning the resulting dried sludge. Therefore, considering the recovery of energy from both the vapour/condensate obtained during the MW drying process, as well as from the combustion of dry sludge, the specific energy output may be decreased from approximately 5.5 to 4 MJ kg^{-1} (i.e., from 1.5 to 1.1 kWh kg^{-1}). In addition, the MW drying energy expenditures may be further reduced by employing more efficient mechanical dewatering systems resulting in a sludge to be MW irradiated with a starting higher DS content. As a result, the specific energy output may be considerably reduced, while increasing the energy recovery from combustion of dry sludge. Nonetheless, the elemental characterization of the dried sludge indicated a fraction of C and N of 30.7 and 5.1%, respectively. This may be associated with the formation of nitrogen oxides (NO_x), and carbon dioxide (CO_2) which may have a negative impact on the environment. It may be required, depending on local regulations, to implement additional filters for the treatment of the exhaust gasses formed during the combustion. As such, this may result in an increase in the capital and operational costs needed for running the system. Recovering energy from the generated vapours/condensate alone may not exhibit such inconveniences. In addition, the implementation of renewable energy sources for powering the MW drying system such as photovoltaic cells may also be a possibility (as discussed in more detail in Section 5.3.5).

Table 10: Elemental composition and calorific value of the mechanically dewatered sludge on a dry mass basis

Parameters	Units	Mechanically dewatered sludge
C	%	30.7 ± 2.6
H	%	5.2 ± 1.1
N	%	5.1 ± 0.9
O	%	28.3 ± 1.9
HHV	MJ kg^{-1}	16.4

The results presented in this section, indicated that at the evaluated conditions the DS content of the sludge was increased up to approximately 30% corresponding to a volume reduction of 83%. However, by simply increasing the irradiation exposure time, a DS content of 95% can be achieved as also reported by other authors (Chen et al., 2014; Mawioo et al., 2016b; Kocbek et al., 2020). MW specific energy expenditures of 5.5 MJ kg^{-1} (1.5 kWh kg^{-1}) are needed for drying the sludge (up to 95% DS), comparable to the energy expenditures on conventional sludge dryers (Kocbek et al., 2020). In addition, a reduction on MW specific energy output may be further achieved by employing more efficient mechanical dewatering systems, by using more efficient MW systems, and by optimizing operational parameters such as the MW power/mass ratio, among others (Chen et al., 2014; Mawioo et al., 2016b). The energy expenditures can also be further reduced by recovering the energy from the condensate/vapour and from burning the dried sludge.

The MW irradiation of the sludge produced a vapour stream containing mostly water and some volatile compounds originally present in the sludge (Deng et al., 2009). Such gas stream condensate before reaching the odour control filters; the condensate was directed to the UF reception tank. Such condensate was characterized for evaluating the recovery possibilities for the resource, and the results of the condensate characterization is presented in Table 11. The condensate derived from the sludge drying process is pathogen free. Overall, the values reported in this study are comparable with the condensate characteristics reported in previous studies involving MW drying of SS (Mawioo et al., 2017). However, the average concentration of TN reported in this study of 33.4 mg L^{-1} was lower than reported by Mawioo et al. (2017) of 213 mg L^{-1}. This could be attributed to an increase in the pH in this study, which could affect the speciation of ammonia from NH_4^+ to NH_3. Ammonium in its gaseous phase (NH_3) could have been volatilized during the condensation and/or the sampling period. The composition of the condensate varies considerably depending on the origin of the sludge and on the technology employed for drying the sludge (Deng et al., 2009; Karwowska et al., 2016). For instance, Karwowska et al. (2016) reported COD, NH_4^+, and PO_4^{3-} concentrations in the condensate of conventional thermal drying systems ranging from 109 to 2,240, from 92 to 343, and from 0.5 to 7 mg L^{-1}, respectively.

Table 11: Condensate physical-chemical characteristics

Parameters	Units	Condensate
COD	mg L^{-1}	565.0 ± 7.8
TN	mg L^{-1}	33.4 ± 8.2
NH_4^+	mg L^{-1}	32.7 ± 7.6
TP	mg L^{-1}	1.8 ± 0.1
PO_4^{3-}	mg L^{-1}	<0.6
pH	-	8.8 ± 0.02

5.3.3 Suitability of the MW processed sludge for land application

Table 12 describes the requirements/regulations set by the United States Environmental Protection Agency (USEPA) and by the European Union (EU) for the reuse of the treated sludge for land applications, as well as the results obtained in this study. Table 12 indicates that the sanitized sludge treated using the MW system fulfils the existent standards regarding the concentration of *E. coli* in the USEPA guidelines. In addition, the sludge must also comply with the USEPA regarding the vector attraction requirements and the presence of heavy metals in the treated sludge. With respect to the vector attraction, a 38% reduction on the initial amount of VS is required to fulfil such a requirement (USEPA, 1994). In the EU, standardized regulations for vector attraction requirements have not been established. However, according to the European Environment Agency, a volatile solids/dry solids (VS/DS) ratio below 0.6 is recommended (Bresters et al., 1997). As shown in Table 12, a VS/DS ratio of 0.3 was reported for the dehydrated sludge. However, the MW system was not able to reduce the original VS content of the sludge, as it was also reported by Mawioo et al. (2016a). The total amount of VS remained unchanged before and after the MW drying. However, if a sufficient reduction in the net amount of water is achieved, (so, the potential for microbial growth is reduced), the sludge may be safely applied to land applications. According to USEPA, this can be achieved when drying the material up to a DS content of at least 90%. Those levels of DS concentrations can be achieved by the MW drying system, although in this study the drying was stopped at a DS concentration of approximately 30% after 26 minutes of irradiation exposure time. The potential land application of the sludge may also be limited by the presence of heavy metals such as Hg, Ni, Zn, Pb, Cr, Cd Cu. The concentration of such compounds in this study were all considerably below the USEPA and EU standards as indicated in Table 12. These results ascertain that the sludge can be safely reused in agricultural application; in addition, the nutrients that can also be recovered would exhibit an additional advantage for enhancing crop productivity.

Table 12: Physical-chemical properties for the sanitized sludge compared to the USEPA and EU standards for treated sludge land applications

Parameter		Sanitized sludge	USEPA	EU
Pathogen requirement				
E. coli	cfu gDS^{-1}	n.d.[a]	<1000	-
Vector attraction requirement				
VS/DS	-	0.3 ± 0.1	-	0.6[b]
VS reduction	%	-	38	-
Metal concentration limitation				
Cr	mg kg^{-1}	117.5 ± 36.8	-	-
Zn	mg kg^{-1}	722.2 ± 252.9	2800	4000
Cu	mg kg^{-1}	60.0 ± 31.4	1500	1750
Ni	mg kg^{-1}	15.2 ± 0.2	420	400
Pb	mg kg^{-1}	10.5 ± 0.3	300	1200
Hg	mg kg^{-1}	0.2 ± 0.1	17	25
Cd	mg kg^{-1}	<0.0001	39	40

[a] not detected/below the detection limit
[b] vector attraction reduction recommended value by the European Environmental Agency

5.3.4 Performance of the membrane separation system for the treatment of the sludge filtrate and condensate

5.3.4.1 Membrane separation technology

The filtrate from the mechanical dewatering unit and the condensate from the MW drying unit were collected in the UF tank and filtered through the UF ceramic membrane system. As indicated in the schematic in Figure 48, the volume in the UF tank fluctuated (by the use of a level sensor) between a maximum volume of 130 L (the volume of a batch of SS) and a minimum volume of 40 L (minimum volume required to have the ceramic membrane submerged). Table 13 shows the characterization of the filtrate from the mechanical dewatering unit, the condensate from the MW drying unit, as well as the UF permeate and concentrate from the UF filtration system. The condensate only contributed a minor fraction of the entire flow (0.4 L out of 130 L); thus, the contribution of the condensate could be neglected.

The UF ceramic membranes had a pore size of 0.08 μm; thus, only the removal of particles larger than the pore size of the membrane was expected. This included bacteria such as *E. coli* which were not detected in the UF permeate (concentration reduced below the detection limit). However, marginal differences were noted in the DS content of the filtrate compared to the DS content in the UF permeate and concentrate, confirming that most of the DS is in the dissolved form (as discussed in Section 5.3.1). Furthermore, as shown in Table 13, most of the soluble compounds such as COD, NH$_4^+$ and K$^+$, passed through the UF filtration. The mass balance carried out in Figure 48 also confirmed that observation; the mass of the soluble compounds could be directly related to the volume of either the UF permeate or the UF concentrate produced during the UF filtration (i.e., 70% of COD, NH$_4^+$ and K$^+$ mass for a particular batch remained in the permeate, while the 30% remaining remained in the UF concentrate following the permeate/concentrate volume ratio). Moreover, Table 13 and Figure 48 indicate that both the total phosphorus and orthophosphate were removed from the sludge filtrate, but were present neither in the permeate nor in the concentrate. Ferric chloride was added to destabilize the particles in the mechanical dewatering process; therefore, some phosphate precipitation could have occurred in the UF concentrate. When opening the UF tank for maintenance interventions, brown/red precipitates in the pipes, surface of the tank, and in the surface of the UF ceramic membrane were observed, indicating the formation of such precipitates. Thus, some of the TP could remain in such precipitates and not detected when sampling.

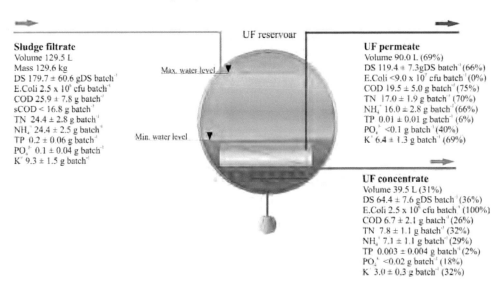

Sludge filtrate
Volume 129.5 L
Mass 129.6 kg
DS 179.7 ± 60.6 gDS batch[-1]
E.Coli 2.5 x 10[8] cfu batch[-1]
COD 25.9 ± 7.8 g batch[-1]
sCOD < 16.8 g batch[-1]
TN 24.4 ± 2.8 g batch[-1]
NH$_4^+$ 24.4 ± 2.5 g batch[-1]
TP 0.2 ± 0.06 g batch[-1]
PO$_4^{3-}$ 0.1 ± 0.04 g batch[-1]
K[+] 9.3 ± 1.5 g batch[-1]

UF reservoar

Max. water level

Min. water level

UF permeate
Volume 90.0 L (69%)
DS 119.4 ± 7.3gDS batch[-1] (66%)
E.Coli <9.0 x 10[7] cfu batch[-1] (0%)
COD 19.5 ± 5.0 g batch[-1] (75%)
TN 17.0 ± 1.9 g batch[-1] (70%)
NH$_4^+$ 16.0 ± 2.8 g batch[-1] (66%)
TP 0.01 ± 0.01 g batch[-1] (6%)
PO$_4^{3-}$ <0.1 g batch[-1] (40%)
K[+] 6.4 ± 1.3 g batch[-1] (69%)

UF concentrate
Volume 39.5 L (31%)
DS 64.4 ± 7.6 gDS batch[-1] (36%)
E.Coli 2.5 x 10[8] cfu batch[-1] (100%)
COD 6.7 ± 2.1 g batch[-1] (26%)
TN 7.8 ± 1.1 g batch[-1] (32%)
NH$_4^+$ 7.1 ± 1.1 g batch[-1] (29%)
TP 0.003 ± 0.004 g batch[-1] (2%)
PO$_4^{3-}$ <0.02 g batch[-1] (18%)
K[+] 3.0 ± 0.3 g batch[-1] (32%)

Figure 48: Mass balance of the UF filtration system

Table 13: Physical-chemical characteristic of the SS filtrate, condensate, and UF permeate and concentrate

Parameters	Units	Filtrate	Condensate	UF permeate	UF concentrate
DS	%	0.14 ± 0.05	-	0.13 ± 0.01	0.16 ± 0.02
E. coli	CFU mL[-1]	1.9 x 10[1a]	-	<1.0 x 10[-2]	6.4 x 10[1a]
COD	mg L[-1]	200.0 ± 60.0	565.0 ± 7.8	216.7 ± 55.1	170.0 ± 52.9
sCOD	mg L[-1]	<130.0	-	-	-
TN	mg L[-1]	188.7 ± 21.6	33.4 ± 8.2	188.7 ± 21.4	196.7 ± 28.9
NH$_4^+$	mg L[-1]	188.3 ± 19.5	32.7 ± 7.6	178.2 ± 31.4	180.3 ± 27.5
TP	mg L[-1]	1.3 ± 0.5	1.8 ± 0.1	0.1 ± 0.03	0.1 ± 0.1
PO$_4^{3-}$	mg L[-1]	1.0 ± 0.3	<0.6	<0.6	<0.6
K[+]	mg L[-1]	72.0 ± 11.3	-	71.3 ± 14.1	75.7 ± 7.8
pH	-	7.3 ± 0.4	8.8 ± 0.02	7.1 ± 0.2	7.2 ± 0.1
EC	µS cm[-1]	(3.1 ± 0.2) x 10[3]	-	(2.9 ± 0.3) x 10[3]	(3.0 ± 0.1) x 10[3]

[a] calculated from the mass balance

The results of this study indicated that the UF filtration has been able to achieve a high removal efficiency for suspended solids and *E. coli,* while allowing the passage of dissolved components such as nutrients, as well as other salts and minerals present in the filtrate. So, this introduces a severe limitation for the reuse of the UF permeate in irrigation, since the permeate exceeds the maximum allowed concentration for TDS and NH_4^+ stated by the Jordanian standard 893/2006 (Ulimat, 2012). Therefore, further treatment of the permeate is needed, if the water is intended to be used for irrigation applications. Eventually, the UF filtration may be proposed as a pre-treatment process for RO systems as discussed next.

The UF permeate was collected in an RO reservoir from which it was directed to a RO filtration unit. The physical-chemical characteristics of the UF permeate (influent to the RO system), RO permeate, and RO concentrate are presented in Table 14. The mass balance of the RO filtration process for a SS batch is presented in Figure 49. The RO filtration process produces both an RO permeate containing a low concentration of ions, and an RO concentrate consisting primarily of the rejected salts and minerals. This is confirmed both in Table 14 and Figure 49. As an example of the removal capacities of ions exhibited by the RO filtration system (as illustrated in Figure 49), the amount of K^+ in the RO permeate (0.1 g) was much lower than both in the UF permeate (influent to the RO system of 6.2 g), and in the RO concentrate (6.4 g), indicating average removal efficiencies of 98%. Table 14 and Figure 49 also indicate good removal efficiencies for the rest of the evaluated compounds. The absence of *E. coli* both in the RO permeate and concentrate was also confirmed. Furthermore, the RO permeate exhibited a low EC of approximately 121 μS cm^{-1} producing a suitable water for reuse in irrigation; in addition, the RO permeate can be also mixed with groundwater enhancing the amount of water resources for irrigation.

The TDS removed from the UF permeate were concentrated in the RO concentrate by a factor of approximately two. For instance, the concentration of K^+ increased from 71.3 in the RO feed to 156.0 mg L^{-1} in the RO concentrate. The RO concentrate exhibited high concentration of nutrients which can be subsequently recovered by applying several treatment alternatives including chemical precipitation methods, and/or for instance stripping processes particularly to recover NH_4^+. Furthermore, the energy expenditures of the UF and RO units for treating one batch of SS were 0.55 kWh and 0.06 kWh, respectively. During the research period, membrane fouling was not observed. However, RO fouling is anticipated to occur over an extended treatment period, which will lead to an increase in both operational and maintenance costs.

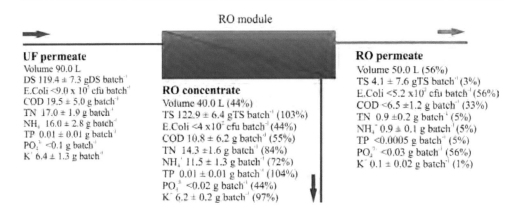

RO module

UF permeate
Volume 90.0 L
DS 119.4 ± 7.3 gDS batch⁻¹
E.Coli <9.0 x 10⁵ cfu batch⁻¹
COD 19.5 ± 5.0 g batch⁻¹
TN 17.0 ± 1.9 g batch⁻¹
NH₄ 16.0 ± 2.8 g batch⁻¹
TP 0.01 ± 0.01 g batch⁻¹
PO₄³⁻ <0.1 g batch⁻¹
K⁺ 6.4 ± 1.3 g batch⁻¹

RO concentrate
Volume 40.0 L (44%)
TS 122.9 ± 6.4 gTS batch⁻¹ (103%)
E.Coli <4 x10⁵ cfu batch⁻¹ (44%)
COD 10.8 ± 6.2 g batch⁻¹ (55%)
TN 14.3 ±1.6 g batch⁻¹ (84%)
NH₄ 11.5 ± 1.3 g batch⁻¹ (72%)
TP 0.01 ± 0.01 g batch⁻¹ (104%)
PO₄³⁻ <0.02 g batch⁻¹ (44%)
K⁻ 6.2 ± 0.2 g batch⁻¹ (97%)

RO permeate
Volume 50.0 L (56%)
TS 4.1 ± 7.6 gTS batch⁻¹ (3%)
E.Coli <5.2 x10⁵ cfu batch⁻¹ (56%)
COD <6.5 ±1.2 g batch⁻¹ (33%)
TN 0.9 ±0.2 g batch⁻¹ (5%)
NH₄ 0.9 ± 0.1 g batch⁻¹ (5%)
TP <0.0005 g batch⁻¹ (5%)
PO₄³⁻ <0.03 g batch⁻¹ (56%)
K⁻ 0.1 ± 0.02 g batch⁻¹ (1%)

Figure 49: Mass balance of the reverse osmosis system

Table 14: Physical-chemical characteristic of the UF and RO permeate and RO concentrate

Parameters	Units	UF permeate	RO permeate	RO concentrate
DS	%	0.13 ± 0.01	<0.06	0.31 ± 0.002
E. coli	CFU mL⁻¹	<1 x 10⁻²	<1 x 10⁻²	<1 x 10⁻²
COD	mg L⁻¹	216.7 ± 55.1	<130.0	270 ± 155.6
TN	mg L⁻¹	188.7 ± 21.4	17.1 ± 3.4	357.7 ± 39.6
NH₄⁺	mg L⁻¹	178.2 ± 31.4	17.0 ± 2.6	287.5 ± 31.8
TP	mg L⁻¹	0.1 ± 0.03	<0.009	0.2 ± 0.1
PO₄³⁻	mg L⁻¹	<0.6	<0.6	<0.6
K⁺	mg L⁻¹	71.3 ± 14.1	1.5 ± 0.4	156.0 ± 5.7
pH	-	7.1 ± 0.2	-	8.6 ± 0.1
EC	µS cm⁻¹	(2.9 ± 0.3) x 10³	(0.12 ± 0.029) x 10³	(5.5 ± 0.5) x 10³

The application of the UF filtration followed by the RO filtration resulted in an efficient method for removing both suspended and dissolved constituents originally contained in the filtrate of the SS after going through a mechanical dewatering process. In assessing the overall treatment performance of the filtration system, treatment efficiencies of over 95% were reported for *E. coli*, DS, TN, TP, NH₄⁺, and K⁺, and of approximately 64% for COD and PO₄³⁻. By following such approach, the filtrate from the SS after mechanical dewatering could be converted into clean water suitable for several reclamation applications such as crop irrigation, and process water, among others; in addition, the RO concentrate can be further treated to recover the nutrients contained in the concentrate at relatively high concentrations.

5.3.5 Outlook on the use of integrated technologies in the treatment of septic tank sludge

The mechanical dewatering unit was quite successful in increasing the DS content of the SS from approximately 0.16% to 5.6%. This had a positive impact on reducing the subsequent MW drying needs (in terms of equipment and energy expenditure) to sanitise and dry the SS. In addition, the performance of the MW drying (energy expenditure) can be enhanced by employing more efficient MW generators, optimising the MW power output/sludge mass ratio, and by recuperating the energy from both the condensate, and from combusting the dried sludge. However, the exhaust gasses formed during the combustion of the dried sludge must be monitored and treated if they represent a source of air pollution due to the formation of NOx and CO_2 gasses.

Furthermore, the UF and RO membrane filtration produced water that could be reused. The SS treated in this study exhibited a high concentration of TDS. Such compounds could have been present in the SS due to the brackish groundwater intrusion in the septic tanks that is commonly observed in the Jordan area (Halalsheh 2008). The reuse of water for irrigation, therefore, would definitely require combining UF and RO filtration in a series. In the absence of brackish groundwater intrusion, combining UF and RO filtration in a series may also be justified when treating SS with a high concentration of ions such as NH_4^+ and PO_4^{3-} – dependent on the local water reuse legislation (Forbis-Stokes et al. 2021). In addition, from the RO concentrate, resources such as nutrients may be recovered by either precipitation or stripping. The results from this study indicate that the evaluated combination of technologies was effective on the sanitisation and drying of the SS, while simultaneously recovering resources such as water for irrigation applications.

The energy demand of the evaluated system was also estimated considering the power demand and operational time of each component when treating one batch of SS (defined as a 130 L of SS). The total energy demand per batch was estimated at 1.8 kWh. Therefore, the energy demand for a hypothetical sludge treatment system with a treatment capacity of 10 m^3 of SS per day (i.e., the treatment capacity of the SS collected by a standard 10 m^3 sludge truck per day) can be extrapolated at approximately 134.1 kWh-day. The MW drying unit (including the water-cooling system) would consume approximately 56% of that energy, the membrane filtration system (UF and RO) consumes 35%, and the mechanical dewatering unit (including coagulation and flocculation) consumption is 8%. The sludge treatment plant should be designed for being deployed in the operational site with minimum interventions. So, the system shall include possibilities for a direct grid connection, in case such possibility is offered at the operational/deployment site. However, off-grid possibilities should also be provided such as the option for powering the system with a diesel/petrol generator or an off-grid photovoltaic (PV) system. Considering the high solar irradiance in Jordan (where the system was evaluated), the energy requirements can be supplied by a PV system (i.e., a renewable source of energy). In PV systems, the specific energy yield, or simply the yield, describes the energy output of the PV plant over one full year and normalized to the peak power ratings of the plant, that is kWh kWp^{-1}. The specific energy yield, or the energy produced for every kW peak (kWp), of a PV module over the course of a typical year in Jordan is approximately 1,600 kWh kWp^{-1} (Al-Addous et al., 2017). The treatment of one sludge truck per day (10 m^3 day^{-1}) would require an energy expenditure of approximately 134.1 kWh-day (approximately 50,000 kWh-year). A PV system delivers 0.10 kWp per square meter. Approximately 31.25 kWp would be needed to power such a system, translated into a PV system of approximately 312.5 m^2. Given that the area needed to accommodate the PV modules is widely available in Jordan, it would be technically feasible to meet the system energy demand by means of renewable sources of energy. Two options are envisioned: (i) a non-mobile sludge treatment system can be constructed in a particular location for a specific localized application, and the PV system can be built in such location; or (ii) a mobile MSD sludge treatment unit can be projected, and the mobile system can be operated with batteries away from the PV energy generation site. The PV modules can generate direct current (DC), and due to intermittency of solar supply, they can be combined with an electricity storage system (i.e., batteries). In this sense, the entire energy demand needed to operate the sludge treatment system is transferred through the batteries. The charged batteries can be placed in the MSD sludge treatment system, and the system can be deployed in the field. The batteries once consumed can be replaced with fully charged batteries, and in such a way secure the operation of the treatment system. As such, the MSD system can be potentially operated in a sustainable manner; thus, reducing the carbon footprint of the overall system operation. This is especially relevant for applications in Jordan, since there is plenty of solar radiation all year around, and also Jordan cannot rely on its own supply of fossil fuels for electricity generation and imports approximately 96% of the energy to meet the local needs (Azzuni et al., 2020). The treatment capacity of the evaluated pilot system can easily be increased to fit such a treatment capacity of 10 m^3 of SS per day. If larger treatment capacities are needed, several containers can be provided in parallel. However, for large treatment demands in the context of centralized sanitation systems, a larger unit would need to be constructed on the site and without conserving the movable features.

This work indicates that the proposed combination of technologies does enable sludge sterilization and drying, while also providing resource recovery. In addition, the outcomes of this study reveal that MSD-based applications, with designs that enable a quick distribution and installation, may contribute to the provision of sanitation in decentralized and non-sewer contexts (as in emergency situations) that show a rapid generation of sludge. Nevertheless, further improvements are needed, which include: (i) optimizing the overall energy consumption of the MSD system; (ii) improving the efficacy of the mechanical dewatering process; (iii) reducing the fouling potential of the membrane filtration system; and (iv) enhancing the MW unit performance. Moreover, the concept should be evaluated by treating various types of sludge, including fresh FS from pit latrines and other onsite sanitation facilities from diverse locations.

Furthermore, as pointed out by Mawioo et al. (2017), for particular applications – such as in the early stages of sanitation provision after the occurrence of an emergency situation – there are very few sludge treatment options available for treating large amounts of fresh sludge generated over such a short period of time. In situations of this nature, it is important to give priority to the inactivation of pathogens for reducing the sludge moisture content to prevent the propagation of excreta-derived endemics. The sanitization approach requires much lower exposure times when compared to sludge drying (i.e., a much lower energy demand). These sorts of scenarios are ideal for the application of the technology evaluated in this study.

5.4 Conclusion

- Three different individual technologies (a mechanical dewatering process, a MW drying system, and a membrane separation process) were successfully integrated for the treatment of SS in a real application in the Jordan Valley in Jordan. The treatment of SS involved the sanitization and dehydration of the SS, while simultaneously producing (recovering) resources such as energy, water, and nutrients.

- The mechanical dewatering processes removed approximately 99.6% of the initial SS water content concentrating the SS from 0.16 to 5.6 DS content with a positive impact in the subsequent MW drying process.

- The MW drying system removed *E. coli* below the detection limit and dehydrated the sludge up to a DS content higher than 95% at energy expenditures of 0.75 MJ kg^{-1} (0.2 kWh kg^{-1}) and 5.5 MJ kg^{-1} (1.5 kWh kg^{-1}), respectively.

- The energy expenditures of the MW drying system can be reduced by 10 and 30% by recovering the energy in the condensate and by combusting the dried sludge, respectively

- The MW dried sludge is suitable for land application exhibiting a VS/DS ration below 0.6 and a DS concentre higher than 90%; in addition, the dried sludge exhibited heavy metals concentrations below the standards for land application

- The combination of UF and RO filtration systems in series produced a high quality permeate ideal for water reclamation in irrigation applications

- Due to the geographical location of Jordan, high solar irradiation was observed, then PV systems can be applied for providing the energy requirements of the evaluated system.

References

Afolabi OO, Sohail M. (2017). Microwaving human faecal sludge as a viable sanitation technology option for treatment and value recovery–A critical review. Journal of Environmental Management, 187, 401-415.

Al-Addous M, Dalala Z, Class CB, Alawneh F, Al-Taani H. (2017). Performance analysis of off-grid PV systems in the Jordan Valley. Renewable Energy, 113, 930-941.

Azzuni A, Aghahosseini A, Ram M, Bogdanov D, Caldera U, Breyer CJS. (2020). Energy Security Analysis for a 100% Renewable Energy Transition in Jordan by 2050. Sustainability 2020, 12(12), 4921.

Banik S, Bandyopadhyay S, Ganguly SJBt. (2003). Bioeffects of microwave—a brief review. 87(2), 155-159.

Bennamoun L, Chen Z, Afzal MT. (2016). Microwave drying of wastewater sludge: Experimental and modeling study. Drying Technology, 34(2), 235-243.

Bresters A, Coulomb I, Deak B, Matter B, Saabye A, Spinosa L, Utvik A, Uhre L, Meozzi P. (1997). Sludge treatment and disposal. EEA, Environmental Issues Series, 7.

Chen Z, Afzal MT, Salema AA. (2014). Microwave drying of wastewater sewage sludge. Journal of Clean Energy Technologies, 2(3), 282-286.

Das D, Baitalik S, Haldar B, Saha R, Kayal NJJoPM. (2018). Preparation and characterization of macroporous SiC ceramic membrane for treatment of waste water. 25(4), 1183-1193.

Deng W-Y, Yan J-H, Li X-D, Wang F, Zhu X-W, Lu S-Y, Cen K-F. (2009). Emission characteristics of volatile compounds during sludges drying process. Journal of Hazardous Materials, 162(1), 186-192.

Englund M, Carbajal JP, Ferré A, Bassan M, Hoai Vu AT, Nguyen V-A, Strande L. (2020). Modelling quantities and qualities (Q&Q) of faecal sludge in Hanoi, Vietnam and Kampala, Uganda for improved management solutions. Journal of Environmental Management, 261, 110202.

Evans NG, Hamlyn MG. (1996). Microwave Firing At 915 MHz - Efficiency and Implications. MRS Proceedings, 430, 9.

Forbis-Stokes AA, Kalimuthu A, Ravindran J, Deshusses MA. (2021). Technical evaluation and optimization of a mobile septage treatment unit. Journal of Environmental Management, 277, 111361.

Friedl A, Padouvas E, Rotter H, Varmuza K. (2005). Prediction of heating values of biomass fuel from elemental composition. Analytica Chimica Acta, 544(1), 191-198.

Gold M, Harada H, Therrien J-D, Nishida T, Cunningham M, Semiyaga S, Fujii S, Dorea C, Nguyen V-A, Strande LJEt. (2018). Cross-country analysis of faecal sludge dewatering. 39(23), 3077-3087.

Günther I, Horst A, Lüthi C, Mosler H-J, Niwagaba CB, Tumwebaze IK. (2011). Where do Kampala's poor "go"?-Urban sanitation conditions in Kampala's low-income areas.

Halalsheh M. Characterization of septage discharging to Khirbit As Samra treatment plant. University of Jordan, Amman-Jordan, 2008.

Hong SM, Park JK, Lee Y. (2004a). Mechanisms of microwave irradiation involved in the destruction of fecal coliforms from biosolids. Water Research, 38(6), 1615-1625.

Hong SM, Park JK, Lee YJWR. (2004b). Mechanisms of microwave irradiation involved in the destruction of fecal coliforms from biosolids. 38(6), 1615-1625.

Hong SM, Park JK, Teeradej N, Lee Y, Cho Y, Park C. (2006). Pretreatment of sludge with microwaves for pathogen destruction and improved anaerobic digestion performance. Water Environment Research, 78(1), 76-83.

Hong Y-d, Lin B-q, Li H, Dai H-m, Zhu C-j, Yao H. (2016). Three-dimensional simulation of microwave heating coal sample with varying parameters. Applied Thermal Engineering, 93, 1145-1154.

Hube S, Eskafi M, Hrafnkelsdóttir KF, Bjarnadóttir B, Bjarnadóttir MÁ, Axelsdóttir S, Wu B. (2020). Direct membrane filtration for wastewater treatment and resource recovery: A review. Science of The Total Environment, 710, 136375.

Ingallinella A, Sanguinetti G, Koottatep T, Montangero A, Strauss M. (2002). The challenge of faecal sludge management in urban areas-strategies, regulations and treatment options. Water Science and Technology, 46(10), 285-294.

Jakobsson C. (2014). Sustainable agriculture: Baltic University Press.

Jiménez B, Drechsel P, Koné D, Bahri A. (2009). Wastewater, sludge and excreta use in developing countries: an overview. In Wastewater irrigation and health (pp. 29-54): Routledge.

Jiménez B, Drechsel P, Koné D, Bahri A, Raschid-Sally L, Qadir M. (2010). Wastewater, sludge and excreta use in developing countries: an overview. astewater Irrigation, 1.

Jones DA, Lelyveld T, Mavrofidis S, Kingman S, Miles N. (2002). Microwave heating applications in environmental engineering—a review. Resources, conservation and recycling, 34(2), 75-90.

Kacprzak M, Neczaj E, Fijałkowski K, Grobelak A, Grosser A, Worwag M, Rorat A, Brattebo H, Almås Å, Singh BR. (2017). Sewage sludge disposal strategies for sustainable development. Environmental Research, 156, 39-46.

Karlsson M, Carlsson H, Idebro M, Eek C. (2019). Microwave heating as a method to improve sanitation of sewage sludge in wastewater plants. IEEE Access, 7, 142308-142316.

Karwowska B, Sperczyńska E, Wiśniowska E. (2016). Characteristics of reject waters and condensates generated during drying of sewage sludge from selected wastewater treatment plants. Desalination and water treatment, 57(3), 1176-1183.

Kocbek E, Garcia HA, Hooijmans CM, Mijatović I, Lah B, Brdjanovic D. (2020). Microwave treatment of municipal sewage sludge: Evaluation of the drying performance and energy demand of a pilot-scale microwave drying system. Science of The Total Environment, 742, 140541.

Kramer FC, Shang R, Heijman SG, Scherrenberg SM, van Lier JB, Rietveld LC. (2015). Direct water reclamation from sewage using ceramic tight ultra-and nanofiltration. Separation and Purification Technology, 147, 329-336.

Kramer FC, Shang R, Rietveld LC, Heijman SJG. (2020). Fouling control in ceramic nanofiltration membranes during municipal sewage treatment. Separation and Purification Technology, 237, 116373.

Kulabako RN, Nalubega M, Wozei E, Thunvik R. (2010). Environmental health practices, constraints and possible interventions in peri-urban settlements in developing countries–a review of Kampala, Uganda. International journal of environmental health research, 20(4), 231-257.

Kumar C, Joardder M, Farrell TW, Karim M. (2016). Multiphase porous media model for intermittent microwave convective drying (IMCD) of food. International Journal of Thermal Sciences, 104, 304-314.

Léonard A, Vandevenne P, Salmon T, Marchot P, Crine M. (2004). Wastewater Sludge Convective Drying: Influence of Sludge Origin. Environmental Technology, 25(9), 1051-1057.

Mawioo PM, Garcia HA, Hooijmans CM, Velkushanova K, Simonič M, Mijatović I, Brdjanovic D. (2017). A pilot-scale microwave technology for sludge sanitization and drying. Science of the Total Environment, 601, 1437-1448.

Mawioo PM, Hooijmans CM, Garcia HA, Brdjanovic D. (2016a). Microwave treatment of faecal sludge from intensively used toilets in the slums of Nairobi, Kenya. Journal of Environmental Management, 184, 575-584.

Mawioo PM, Rweyemamu A, Garcia HA, Hooijmans CM, Brdjanovic D. (2016b). Evaluation of a microwave based reactor for the treatment of blackwater sludge. Science of the Total Environment, 548, 72-81.

Murungi C, van Dijk MP. (2014). Emptying, transportation and disposal of feacal sludge in informal settlements of Kampala Uganda: the economics of sanitation. Habitat international, 42, 69-75.

Nikiema J, Cofie OO. (2014). Technological options for safe resource recovery from fecal sludge.

Peal A, Evans B, Blackett I, Hawkins P, Heymans C. (2014). Fecal sludge management: a comparative analysis of 12 cities. Journal of Water, Sanitation and Hygiene for Development, 4(4), 563-575.

Pino-Jelcic SA, Hong SM, Park JK. (2006). Enhanced anaerobic biodegradability and inactivation of fecal coliforms and Salmonella spp. in wastewater sludge by using microwaves. Water environment research, 78(2), 209-216.

Radford J, Sugden S. (2014). Measurement of faecal sludge in-situ shear strength and density. Water SA, 40(1), 183-188.

Ronteltap M, Dodane P-H, Bassan MJFSM-SAI, Operation. IWA Publishing L, UK. (2014). Overview of treatment technologies. 97-120.

Rose C, Parker A, Jefferson B, Cartmell E. (2015). The Characterization of Feces and Urine: A Review of the Literature to Inform Advanced Treatment Technology. Critical Reviews in Environmental Science and Technology, 45(17), 1827-1879.

Rusydi AF. (2018). Correlation between conductivity and total dissolved solid in various type of water: A review. Paper presented at the IOP Conference Series: Earth and Environmental Science.

Schaum C, Lux J. (2010). Sewage sludge dewatering and drying. ReSource–Abfall, Rohstoff, Energie, 1, 727-737.

Septien S, Singh A, Mirara SW, Teba L, Velkushanova K, Buckley CA. (2018). 'LaDePa' process for the drying and pasteurization of faecal sludge from VIP latrines using infrared radiation. South African journal of chemical engineering, 25, 147-158.

Strande L, Schoebitz L, Bischoff F, Ddiba D, Okello F, Englund M, Ward BJ, Niwagaba CB. (2018). Methods to reliably estimate faecal sludge quantities and qualities for the design of treatment technologies and management solutions. Journal of Environmental Management, 223, 898-907.

Stuerga D. (2006). Microwave-material interactions and dielectric properties, key ingredients for mastery of chemical microwave processes (Vol. 2): WILEY-VCH Verlag GmbH & Co. KGaA.

Sykes GB, Skinner FA. (2015). Microbial aspects of pollution: Elsevier.

Thye YP, Templeton MR, Ali MJCRiES, Technology. (2011). A critical review of technologies for pit latrine emptying in developing countries. 41(20), 1793-1819.

Ulimat A. (2012). Wastewater production, treatment, and use in Jordan. Paper presented at the Second Regional Workshop Safe Use of Wastewater in Agriculture, New Delhi, India.

USEPA USEPA. (1994). A Plain English Guide to the EPA Part 503 Biosolids Rule: US Environmental Protection Agency.

Vadivambal R, Jayas D. (2010). Non-uniform temperature distribution during microwave heating of food materials—A review. Food and bioprocess technology, 3(2), 161-171.

Wei Y, Van Houten RT, Borger AR, Eikelboom DH, Fan Y. (2003). Minimization of excess sludge production for biological wastewater treatment. Water Research, 37(18), 4453-4467.

6

Reflections and outlook

6.1 Reflections

This thesis addressed several research gaps in the field of sludge treatment by microwave technology. In the following paragraphs, reflection on the progress made concerning the main research questions is given. As it is often the case in PhD studies, my PhD research answered many questions that were posed at the start but also opened even more questions at the end as my horizons and fundamental understanding of the topic increased. My reflection on the four main interests of my research is presented below.

6.1.1 On the status of the economic potential of microwave technology in the treatment of sludge from the standpoint of operational costs

The overall specific energy consumption reported in microwave treatment of sludge can be reduced by 70% in comparison to present studies, and this can be achieved by improving the design and construction of the microwave system, including the drying chamber, the microwave generator and the water vapour extraction unit. Once this is in place, the specific energy consumption of the system will be in the range of traditional drying technologies (e.g., approximately 1.0 kWh L^{-1} or 3.6 MJ L^{-1}).

The first hypothesis, as shown above, was posed with reference to specific energy consumption (SEC) of the microwave sludge drying process. Within the scope of the study, the emphasis was placed towards the design features incorporated in the novel microwave-based pilot-scale unit that may directly affect the energy performance of the system. For instance, the volume of vapours produced in a unit of time and the implementation of water vapour extraction techniques with an installed capacity that was able to remove vapours efficiently from the drying chambers represents a fundamental improvement that had a major impact in the performance of the system. The latter prevented condensation and subsequent rewetting of the sample during the drying process, experienced in previous studies (Mawioo et al., 2017). At the same time, the uniformity of the electromagnetic energy distribution within the applicator and sample was addressed through the introduction of a turntable and by dimensioning/sizing the drying chamber. Further as described in this dissertation, the specific energy consumption can be reduced by optimising the operational parameters, specifically microwave power output, which had a substantial impact on the conversion efficiency of the microwave generation efficiency and, thus on the improvements in the energy performance of the system. The combined effects of the incorporated technical features in the system and optimisation of operational parameters, therefore, proved to be an appropriate strategy in achieving an increase in energy efficiency, with a corresponding decrease on the specific energy consumption, which was approximately 4.3 MJ L^{-1} (i.e., 1.2 kWh L^{-1}). This result means that the reported specific energy consumption experimentally determined falls within the range of traditional drying techniques. It was identified that further reduction in specific energy consumption of a microwaves dryer might be achieved by; i) introduction of heat exchanger for the recuperation of energy from vapours/condensate and ii) utilisation of industrial microwave systems with higher quoted efficiency. All of which may ultimately reduce the energy consumption of the system to approximately 2.5 MJ L^{-1} (i.e., 0.7 kWh L^{-1}), thus making it competitive technology to traditional drying approaches such as convective and conductive dryers. In addition, renewable energy sources could be used to meet the electrical energy needs of the system.

Furthermore, in the assessment of the system performance, it has been observed that there is a large gap in the knowledge of the electromagnetic distribution within the reaction chamber and the sample. This is attributed to the absence of measuring instruments that can enable direct quantification of the distribution and strength within the microwave chamber and the samples. Consequently, the effects of the design features implemented in the system, which include a change in the design of the drying chamber, number and position of the waveguide, and others on the drying and energy performance of the system, still remain partially understood. This highlights the importance of developing a tool that may provide a better understanding of the technical aspect of the proposed design solutions on its effect on the propagation of electromagnetic energy within the system. Using this information, a more efficient microwave system can be designed and thus the energy performance of the system optimised. A proposed method to address the latter is found in Section 6.2.1 of this chapter.

6.1.2 On the status of moisture levelling effect on the limitation imposed by the microwave penetration depth

The thickness of the sample limits the propagation of (microwave-induced) energy within the sample, and this limitation can be counteracted by the microwave-selective 'moisture levelling' effect.

The second hypothesis concerned the operational parameters, such as the thickness level, and was raised with reference to the so-called 'moisture levelling effect', which despite being repeatedly cited in the literature as an important advantage of microwave heating, has not been actually well understood. This effect is known to enhance the microwave power absorption in sample parts with higher moisture content; hence offsetting the uneven heating caused by the constraints associated with the limitations imposed by the levels to which the microwave energy can penetrate the material (i.e., in sludge materials the microwave penetration depth at a frequency 2,450 MHz and 20°C may vary from 5 to 14 mm).

The method chosen in this study proved to be satisfactory in establishing and confirming the moisture levelling effect, which was determined by analysing the changes in sludge moisture content and the temperature measured along the axial axis of the sample at different sample thicknesses. Specifically, in the evaluated range of sludge thickness (from 45 to 150 mm), it was concluded that the increase in the sample thickness negatively impacts the sludge drying rates, the exposure time and the SEC. However, the changes in the system throughput capacity (±2.6 minutes) as affected by the limitation of the microwave penetration depth may be considered marginal, especially in comparison to conventional drying applications in sludge treatment, where it was observed that the rate of water evaporation is inversely proportional to the thickness of the material.

These results provide important findings on microwave sludge drying that may have a positive effect on their application in areas with limited available space, which is more prevalent in densely populated areas where treatment space competes with other demands.

Although the microwave technology is already recognised for its low physical footprint, the need for small drying surface requirement implies that the drying equipment could be further reduced in size, owing to the moisture levelling effect. Therefore, these results outline the benefits of providing effective and fast methods for sludge volume reduction through the use of microwave technology and thereby fulfilling the objective of this research.

However, this study provided only a stepping stone for further understanding through research on the sludge moisture levelling effect, which information/knowledge, as mentioned previously, is still lacking in the literature. The method developed and utilised in this study to assess the moisture levelling effect proved acceptable for the purpose of this research and is recommended to be reproduced in future studies.

Further research is needed to explore alternative methods that may be used to assess and confirm the moisture levelling effect with respect to the uniformity of the electric field strength within the material. The latter poses a major challenge to researchers, especially as the quantification of moisture levelling is not a straightforward task, owing to the fact that changes in the distribution of the electrical field in the drying chamber are not known in practical experiments. Accordingly, one of the fundamental aspects to consider when evaluating the moisture levelling effect is the electric field variation within the system, which may vary according to the changes in the sludge dielectric properties, and thus the sludge moisture content, temperature, frequency and other parameters. Moreover, it is apparent that many aspects still remain to be studied to deepen understanding of the sludge moisture levelling effect. In addition, dielectric properties of faecal and septic sludge are still not determined with confidence, and this is one of the urgent tasks that have to be addressed in microwave application to drying of sludges, especially when the modelling is concerned. Clearly, the present work offers several opportunities and directions for further research projects on studying factors that may overcome the limitation of microwave penetration depth.

6.1.3 On the status of the effect of water distribution within the sludge matrix on the microwave drying performance

Variation in sludge composition (specifically concerning the distribution of water as the result of its fat and oil content), has a direct influence on the (efficiency of) operation of microwave-based dryers, and thus, on the operational cost and the design of the system.

The third hypothesis was related to the efficiency of microwave absorption by water molecules, which was hypothesised to be dependent on the state of the water as defined by the sorption isotherm, and that its variation is affected by the fat and oil content of the sludge. This hypothesis was confirmed by analysing the disparate data of mass sludge reduction, moisture dependant water activity (sorption isotherms) and the physicochemical properties of sludge used in this research to find out that a correlation exists between the microwave sludge drying performance and the physical binding strength of water molecules within the sludge matrix.

The results obtained from the study confirmed that the microwave energy efficiently heats water molecules such as free water, while the bound water fraction characterised with high binding strength to solids, could not efficiently rotate under the oscillating electromagnetic field, thus hampering the efficiency of the drying process. Further, it was confirmed that the efficiency of heating and vaporisation of water molecules was enhanced, due to the presence of hydrophobic compounds, such as fats and oils, which favoured an increase on the free water content and thereby enhancing the efficiency of the process. The fat and oil content also positively impacted the resistance of the dry product to the weathering effect. Meaning that the material, with lower available sites for absorption of bound water, can be dried to a lower sludge moisture content (i.e., higher DS content) with respect to the available storage conditions; thus, leading to a decrease on the sludge volume/mass and associated sludge transport costs to its final reuse/disposal locations. The novelty of this research, therefore, lies in the produced set of experimental data that provide a fundamental understanding of the effect of water distribution within the sludge matrix on the microwave drying performance. Thus, the applied experiments fulfilled the objective of this research.

However, it was also observed that the microwave drying and energy performance vary according to the compounds that were or were not analysed in this study, such as porosity, permeability, thermal conductivity, organic content, heavy metals, and other compounds forming the sludge matrix, all of which that might interfere with this study to a certain extent, but its effects could not be successfully quantified within the scope of this research. For instance, it is known that the bound water fraction increases with an increase in the organic content of the sludge. These compounds tend to bind water molecules and thus hinder the mobility of water within the sludge matrix, which could ultimately lead to a reduction in sludge drying rates. Many other explanations are offered in Chapter 4, confirming that the sludge composed of a variety of compounds, that might have interfered with the study of fats and oils on the microwave drying performance to a certain degree, but due to multiple interactions between microwave energy and the compound forming the sludge matrix, could not be pinpointed.

The challenge, therefore, lies in the production of a reliable set of experimental data to validate and optimise the influence of the sludge composition on the microwave drying performance. Therefore, despite the fact that the selected, evaluated physicochemical parameters and experimental procedure used performed satisfactorily in this study, there is still a knowledge gap that needs to be explored to provide a further understanding of the interaction between the microwaves and the sludge matrix. This interaction has a substantial influence on the system throughput capacity, and thus on the system design and performance.

6.1.4 On the status of microwave treatment effect on the situ treatment of sludge

Integration of existing technologies for wastewater and sludge treatment, such as microwave technology coupled with mechanical dewatering and membrane separation technology, enables efficient treatment of various types of sludge *in-situ*.

The final hypothesis pertains to the use of microwave-based technology as a possible solution for the rapid *in-situ* treatment of various types of sludge in conjunction with mechanical dewatering units, ultrafiltration unit and reverse osmosis units. It was confirmed that the integrated technologies within the containerised system enable sludge sterilisation, drying, and liquid phase treatment to a level of high-quality water suitable for reuse in a relatively small footprint (i.e., the system was containerised and mounted on a trailer). Alongside the latter, the focus of the research was aimed at proving the concept that strategic use of technological solutions, which designs enable for quick distribution and installation, may improve the sanitation provision in decentralised, non-sewered context. In fulfilling the research objective, however, only one type of sludge was tested (the research was carried out in Jordan and was completed just before COVID-19 outbreak). Hence, in the future, the testing of the system should be extended by treatment of various types of sludges, including fresh faecal sludge from pit latrines and other on-site sanitation facilities from various locations. In addition, during the evaluation, the operational boundaries of the treatment units incorporated in the containerised system, in treatment of septic tank sludge, including the performance of mechanical dewatering unit, ultrafiltration and reverse osmosis unit were not addressed. The lack of which is evident in this study. This was partially attributed to the available time for the testing of the system. That is, the evaluation of the membrane separation technology needs rigorous experimental investigation that should be applied to a different type of sludge over a span of several months to determined its potential for more diverse use in practice (also for other applications, e.g., industrial sludges, sludge from marines and ships etc.). In addition, there is a need for the inclusion of more research on this topic with a focus both on membrane separation technology and mechanical dewatering systems. The latter may ultimately enable further understanding of the operational characteristics of the units, which may directly impact the system efficiency, affordability and potential use of the system in the field.

It should be stressed that the introduction of a technological solution alone has been described as inadequate and as such, many alternative solutions to technical solution need to be considered, including the institutional acceptance, user acceptance, promotion of the system and thus the commercialisation of the system, with the immediate implication in providing a safer environment achieved through minimising the spread of diseases such as diarrheal diseases, cholera, and typhoid, among others. Thus, in the evaluation of the mobile system, many new questions were opened that need further explanation.

6.2 Outlook and recommendations

This thesis introduces the use of the novel microwave-based pilot-scale unit as an alternative technology for the sanitisation and drying of sludge from a municipal wastewater treatment plant and an on-site sanitation facility. The interest in the use of microwave technology in the treatment of sludge is related to the possibility of economic benefits from volumetric heating, moisture levelling effect, increased liquid and vapour migration from the interior to the surface of the product. All of which, as demonstrated in this research lead to faster processing times, enhanced drying rates, and reduced physical footprint requirement. Furthermore, in this research, the concept of integrating the microwave technology with mechanical dewatering process and membrane separation process was successfully applied in *in-situ* treatment of septic tank sludge. The results of this research indicate that operating technologies in the system in a strategic manner may make it possible to treat other types of sludge originating from on-site sanitation facilities, such as public latrine sludge, pit latrines and others. Therefore, possibly alleviating issues related to the management of faecal sludge (collection, transport and treatment) which have been reported to be inadequate and, in some cases, to be completely lacking in both urban and rural areas (Peal et al., 2014). That is, such an integrated system converted waste (sludge) into a sterilised sludge (free of pathogens) allowing for reuse as a fertiliser, and dried sludge with a high calorific value (18-20 MJ/kg) to be used as a fuel if desired. In addition, the system generated various resources, including the production of water for promoting different types of water reuse (depending on the reuse requirements, the system is capable of adjusting the desired level of water treatment). The overall success of the integrated microwave-based pilot plant system developed in this study illustrated the technical and economic potential of such a technology for sludge drying, sanitisation and treatment of liquids which may be potentially used for cases of massive sludge accumulation at confined areas such as in refugee camps. The results of this work indicate high feasibility of the application of microwave-based technology in the protection of the environment and safeguarding public health. Undoubtedly, more information and knowledge are required pertaining to the operation of microwave technology as a commercially available treatment option for sludge.

6.2.1 Fundamental and practical application of microwave in the treatment of sludge

The data obtained from this research provide essential information that was used to partially explain and describe the variations of sludge temperature, mass/volume reduction in relation to the microwave power absorption density, uniformity of microwave energy within the samples and the governing compound determining the absorption of microwave energy and conversion into heat and correspondingly sludge volume reduction. However, the scientific knowledge on sludge drying, describing the interaction between the sludge and microwaves, was only provided at a rather superficial level. The latter is attributed to the lack of instruments available to measure and analyse the distribution and intensity of the electromagnetic field, both in the drying chamber and the sample and are currently being realised by mathematical models or software alone. Thus, in order to generate more fundamental knowledge on these aspects, computer-based modelling and simulation are currently considered as one of the appropriate methods to predict the distribution of electromagnetic energy and intensity within the cavity and the material (sludge). Modelling can improve the understanding of the heating and sludge moisture loss occurring during the drying process. In addition, when the model is used to assess whether the technology is successful, it can speed up and facilitate the development of a better and increasingly efficient tailor-made system for drying of sludge, all the way from simulation to engineering practice, and in a cost-effective way. However, the information required for the description and evaluation of microwave interaction with sludge sample through the use of coupled physics models, require input information and prior knowledge on at least: i) the physicochemical properties of sludge assessed with reference to the dielectric properties of sludge as a function of the drying exposure time, ii) the knowledge on the heat and mass transfer processes and mechanisms occurring during the drying process. The information required for the modelling process may be partially derived from this research. Several other properties of sludge (e.g., porosity, thermal conductivity, permeability and others) should also be taken into account in order to properly examine the drying of sludge through a mathematical modelling study. There are, however, a number of methods available for the analysis of the physicochemical properties of sludge and, as such, the advantages and disadvantages of each method must be assessed, weighted and selected on the basis of a variety of criteria. For instance, several techniques have been established as a standard method to measure the dielectric properties of the material. The methods developed for measuring the dielectric properties include a coaxial probe, free space methods, transmission line and reflection methods, and many more others (Barba et al., 2012; Khan et al., 2012). Each of the aforementioned methods is limited to the desired frequency, required measurement accuracy, temperature, cost and the type of sampled material regarding both physical and electric properties (Barba et al., 2012; Khan et al., 2012). For instance, the open-end coaxial probe, frequently employed for measuring the sludge properties, is a fast method; however, the accuracy of the results is questionable due to sensitivity of apparatus to air gaps and bubbles within the material. Thus, one limitation of the coaxial method is that material can be evaluated under controlled conditions at temperatures below the boiling point of the liquid (Barba et al., 2012). Similarly, methods for evaluating thermal conductivity, specific heat capacity and others may vary substantially in both the approach and the accuracy. The choices, which may ultimately affect the accuracy of the results obtained from the measurements of sludge properties and thus directly influence the success of the mathematical model used for describing the drying process. Therefore, rigorous testing of the material properties with respect to changes subjected to the drying process needs to be carried out, while carefully choosing the methods for evaluation of the physicochemical properties of sludge.

In addition, from an empirical point of view, it is also of interest to elucidate the boundary conditions of the operation condition, specifically the sludge thickness levels with respect to the moisture levelling effect, the temperature distribution within the material and thus the efficiency of the sanitisation of the sludge. In case the treatment goal is pathogen destruction, it is suspected that the thickness levels have a detrimental effect on the temperature sanitisation rate and thus the system throughput capacity, which needs to be evaluated and confirmed in further research. Further consideration should be given to the sanitisation of sludge in terms of both thermal and non-thermal effects. Whereas the author stipulates that the sanitation of the material undergoing microwave irradiation can be mainly attributed to thermal effects and that the non-thermal microwave effects on the sanitation efficiency of the sludge are marginal. In addition, ways of enhancing the moisture levelling effect, or decreasing the limitation imposed by the microwave penetration depth with specific reference to the material being treated may be investigated. For example, the use of microwave generators operated at a frequency of 0.915 GHz can be considered as one of the options that would have a positive impact on the moisture levelling effect due to their microwave penetration depth, which is approximately three times larger than that of the 2.45 GHz generators used in this study.

Furthermore, in the evaluation of the practical application of microwave technology in the treatment of sludge, one major disadvantage associated with the operation of microwave technology was noted; specifically, the cooling water requirements. Microwave system with installed power output capacity up to 2 kW is usually air-cooled, while those with a higher installed power capacity are cooled using water. The microwave system, in case its placed in areas that lack water needed to operate the microwave treatment system, needs to be provided with a cooling system and thus exceeding the specific energy required to increase the sludge dry solid content to 90% with microwave technology alone.

Alternative approaches to providing cooling water required for the operation of the microwave system is the implementation of a heat exchanger, by which the temperature of the cooling water may be reduced while simultaneously recovering the waste heat from the heated water, thus ultimately increasing the efficiency of the drying process. The latter must be supported by calculations with specific reference to condition in which it will be operated and with testing of the system with an installed heat exchanger in real-life application.

Special attention must also be devoted to the capital and maintenance expenditures, which were not assessed in this study. In addition, it must be confirmed that all the elements of the integrated system based on the microwave technology are clean, efficient and environmentally safe technologies. For instance, the exhaust gasses formed during the drying process need to be analysed and measured pollution levels compared with the national statutory guidelines and regulation. Undoubtedly, other research opportunities and questions may be found pertaining to microwave treatment of sludge, all of which may enhance the practical application of microwave technology in the treatment of sludge.

6.2.2 Practical application of microwave in the treatment of sludge in conjunction with other treatment technologies

The results obtained in this study also provided feedback on the operation of mechanical dewatering unit and membrane separation technology in the treatment of septic tank sludge and thus directions for future research as described in the following paragraphs.

The performance of the sludge dewatering unit was not satisfactory when treating raw sludge with low dry solids concentrations (e.g., 0.2%), therefore applied dewatering process needed to be improved to treat the diluted sludge. In addition, problems were observed with the dosing of chemical conditioning reagents, such as ferric chloride and polymers. The dosage needed to be adjusted according to the sludge composition, which variability was a subject of many factors, including the on-site sanitation facilities construction, retention time, rainwater intrusion, and many others. Thus, the inherent variation in faecal and septic tank sludge composition has a significant impact on the dose of the chemical reagents needed to facilitate the mechanical dewatering. Hence, the optimum dosage should be identified each time through the use of, e.g., jar tests with respect to the material processed. In case the chemical reagent's dose is not determined, it may result in reduced efficiency of the dewatering unit and thus the efficiency of the drying process. In addition, in case the polymer dose is exceeded it may lead to a slimy consistency of the filtrate, which may subsequently negatively affect the operation of the subsequent process proposed in the treatment train (i.e., ultrafiltration unit). In all cases of underdosing or overdosing, the cost-effectiveness will be reduced. In addition, ferric chloride aims at stabilising of the sludge particles, precipitates both the particulate and soluble forms of phosphorus, thus increasing the costs of resource-recovery techniques that are used to produce fertiliser (in particular, production of struvite). These observations suggest that the chemical condition agents used may cause substantial problems to the overall operation of the mobile treatment system. Based on the observation, it is therefore recommended to operate the ultrafiltration in direct treatment of screened septic tank sludge, when considering the influent with total dry solids below 1%. In case the solids concentration is larger than 1%, the use of mechanical dewatering devices may be justified. This is a matter that requires further testing. Alternatively, other technological solutions may be integrated within the system boundaries. The choice of which requires a deepen literature review, and perhaps innovations that may enable partial dehydration (i.e., removal of free water) of sludge prior to the drying units and reducing the fouling potential nature of the faecal sludge on the operation of an ultrafiltration unit.

Part of this study dealt with septic tank sludge characterised with low dry solids content (0.16%). Accordingly, the operation of ultrafiltration and reverse osmosis units had a major role in the purification of the septic tank sludge with respect to the impact of its side streams on the environment. As mentioned previously, limited research has been done towards evaluating the operational characteristics of the membrane separation technology, which include, evaluation of the system at various flows, loads, pressures, backwashing frequency using indicators such the water permeability of the membrane, reversible and irreversible fouling, critical flux identification, and many more, that would improve the fundamental understanding of the operation and suitability of the ultrafiltration and reverse osmosis for the treatment of septic tank sludge. In addition, more research is required aiming at the evaluation of the membrane separation technology not only on septic tank sludge but also other types of sludge, such as fresh faecal sludge, which is usually existing in a semi-solid form. In such a case, the incoming stream to the ultrafiltration and reverse osmosis unit (condensate) is namely stemming from the sludge drying process. The results shown in Chapter 5 suggest that the nutrients found in the condensate, namely present in its soluble form, can be directly treated using reverse osmosis unit. However, it is hypothesised that the fouling index that may be evaluated through, for instance, the modified fouling index expressed as in bar per hours is very high, i.e., at least 0.3

bar h^{-1} at a flux of 100 L $m^{-2}h^{-1}$. This value may be translated in the increased chemical cleaning frequency requirement, which as specified by the manufacturer, should be performed when the pressure gradient in a pressure vessel has decreased by approximately 10% or even lower. Thus, the chemical cleaning frequency may be very high and thus potentially results in increased production downtime and decrease the life span of the reverse osmosis membrane. Accordingly, it is also assumed that various forms of fouling may be observed, including pore blocking and cake compression. However, to confirm this hypothesis, several experiments need to be carried out and also to take into account that most of the methods used to predict the rate of fouling in reverse osmosis membranes may lead to an overestimation of the modified fouling index values, silt density index or others. It is recommended that the operation of the reverse osmosis system in the treatment of condensate needs to be evaluated in real-time measurements alone with and without an ultrafiltration unit. Special attention should also be diverted to biological fouling of the membrane. In addition, it is suggested that the microwave throughput capacity and the sludge composition can drastically affect the composition of the condensate, specifically, the concentration of compounds such as chemical oxygen demands, ammonium, and many others. More testing, therefore, needs to be carried out on the evaluation of condensate derived from the microwave treatment system under various operational parameters and the effect of composition on the operation of the subsequent treatment process. In essence, the application of both ultrafiltration unit and reverse osmosis in treatment of condensate is relatively new and therefore requires more research.

In this research, operational issues were observed with the automatization of the system. Specifically, the system should be continuously operated in a way that separates the operator from direct contact with sludge and hence safeguards the operator's health and safety. A possible solution to address the latter, is the introduction of a continuous belt conveyor, pumps and other equipment, as shown in Figure 50. Based on the schematics of the system it may be observed that the configuration of the technological treatment train proposed resembles the current system used in the study, and may enable treatment of sludge with low and high dry solids content. For instance, in case of treatment of sludge produced at a dry solids concentrations from approximately 1 to 4% may be firstly screened, and then passed to an equalisation and conditioning basin. Later on, the sludge is introduced to a coagulation, flocculation and dewatering process. Two streams are produced here: the dewatered sewage/ septic tank sludge/fresh faecal sludge at a dry solids concentration of approximately 15%, and the filtrate. The dewatered sludge is directed to a storage basin from where it is transported to the microwave drying compartment of the prototype. Dried sludge is obtained at a dry solids concentration of approximately 90%. The water vapour produced at the microwave chamber is directed together with the filtrate to the ultrafiltration unit. The energy of the condensation of the water vapour is also used to preheat the sludge in the sludge storage basin. The retentate is returned back to the coagulation, flocculation and dewatering processing, while the filtrate can be either taken as one of the final products of the technology or can be sent to a reverse osmosis process for further polishing. The concentrate from the reverse osmosis may be alternatively passed back to the coagulation, flocculation process, while the permeate is a final product of the technology. Furthermore, in the case of sludge with high dry solids content, the sludge is directly directed to the microwave drying chamber. Whereas, the condensate produced can be directed either to the headworks of the wastewater treatment facility or to the ultrafiltration unit. This way, the prototype is fully automated, operates in a continuous mode (rather than the previous pilot which was operated in a batch mode).

It is recommended to build and test such a system in the future as it may uncap most of the limitations of the integrated system used in this research.

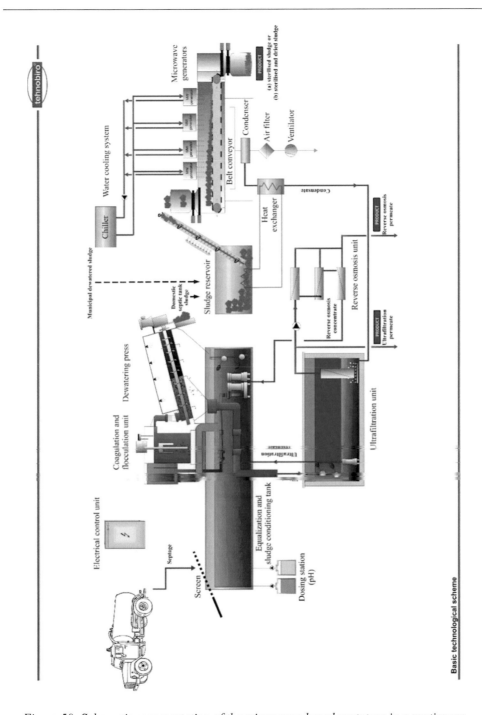

Figure 50: Schematic representation of the microwave-based prototype in a continuous arrangement (Figure acquired from Tehnobiro d.o.o. (Slovenia))

6.2.3 Final remarks

The results of this research showed the potential of the use of microwave-based technology for the treatment of various types of sludges, ranging from sewage sludge to septic and faecal sludge. Achieved specific energy consumption in this study was remarkable 4.3 MJ L^{-1} (i.e., 1.2 kWh L^{-1}) but with some modifications and improvements, the efficiency can be further improved to 2.5 MJ L^{-1} (i.e., 0.7 kWh L^{-1}) which would make this viable technology alternative to commercially available drying techniques.

In this study, the microwave technology was integrated into a system for sludge and water treatment, including odour treatment and energy recovery. The potential of the use of ultrafiltration and reverse osmosis for the treatment of the liquid phase of the sludge and side streams generated in the process becomes obvious in the study. Some improvements should be made in the pre-treatment of sludge, depending on its dry solids content and the future system should have continuous operator-free feeding and handle through the entire process.

Further research is needed to evaluate the environmental effects of an energy recovery unit, especially if obtained through the thermal process.

References

Barba AA, d'Amore MJMmc. (2012). Relevance of dielectric properties in microwave assisted processes. 6, 91-118.

Khan MT, Ali SMJIJoIT, Engineering E. (2012). A brief review of measuring techniques for characterization of dielectric materials. 1(1).

Mawioo PM, Garcia HA, Hooijmans CM, Velkushanova K, Simonič M, Mijatović I, Brdjanovic D. (2017). A pilot-scale microwave technology for sludge sanitization and drying. Science of the Total Environment, 601, 1437-1448.

Peal A, Evans B, Blackett I, Hawkins P, Heymans C. (2014). Fecal sludge management: a comparative analysis of 12 cities. Journal of Water, Sanitation and Hygiene for Development, 4(4), 563-575.

About the Author

Eva Kocbek completed her bachelor's degree in Mechanical Engineering at the University of Maribor, Slovenia in 2014, specializing in Energy, Process, and Environmental Engineering. After graduation, she gained valuable practical experience in engineering and designing solutions for drinking applications with a specialized water company in Slovenia. To extend her working knowledge and understanding of water and sanitation, she later completed a Master of Water Engineering at the UNESCO-IHE Delft Institute for Water Education in Delft, The Netherlands.

During her master's studies, Eva researched ammonium permeability using a reverse osmosis pilot-scale unit treating groundwater for human consumption. Driven by research and innovations, she started her doctoral research as part of a joint program at the Delft University of Technology, Faculty of Applied Science, Department of Biotechnology, and IHE Delft Institute for Water Education (formerly known as UNESCO-IHE), Department of Environmental Engineering and Water Technology. Her research project was developed and funded within the framework of the Programmatic Cooperation Dutch Ministry of Foreign Affairs and the IHE Delft (DUPC2) initiative. The underlying goal of the study was to develop a portable MW-based treatment system for on-site faecal sludge management for the humanitarian and development WASH sector.

She is currently working in the field of R&D, design, and operation of systems to address issues in the area of i) drinking water preparations, ii) the provision of ultra-pure water for the needs of dialysis centers, process water, and iii) sludge treatment.

List of Publications

JOURNAL ARTICLES

Kocbek E, Garcia HA, Hooijmans CM, Mijatović I, Lah B, Brdjanovic D. Microwave treatment of municipal sewage sludge: Evaluation of the drying performance and energy demand of a pilot-scale microwave drying system. Science of The Total Environment 2020; 742: 140541.

Kocbek E, Garcia HA, Hooijmans CM, Mijatović I, Brdjanovic D. Microwave treatment of municipal sewage sludge: Effects of the sludge thickness and sludge mass load on the drying performance. Submitted to Journal of Environmental Management., 2021.

Kocbek E, Garcia HA, Hooijmans CM, Mijatović I, Kržišnik D, Humar M, Brdjanovic D. Effects of the sludge physical-chemical properties on the microwave drying performance of the sludge. Submitted to Journal of Science of The Total Environment, 2021.

Kocbek E, Garcia HA, Hooijmans CM, Mijatović I, Al-Addousd M, Dalala Z, Brdjanovic D. Novel semi-decentralised mobile system for the sanitization and dehydration of septic sludge: A pilot-scale evaluation in the Jordan Valley. Submitted to Journal of Environmental Science and Pollution Research, 2021.

Acknowledgement

I would like to express my heartfelt appreciation and gratitude to my promotor, Prof. Dr. Damir Brdjanović, for granting me an opportunity to work on this one-of-a-kind project and for providing invaluable guidance, support, effort and advice throughout my research. I would like to thank my supervisor, external supervisor, and (co)-promotor Assoc. Prof. Dr. C.M. Hooijmans, Prof. Dr. Ivan Mijatović, and Assoc. Prof. Dr. H.A. Garcia. Thank you for your patience and guidance, as well as for providing constructive suggestions and comments that have greatly improved the quality and content of the research project. Special appreciation deserves Assoc. Prof. Dr. H.A. Garcia, for patiently teaching me how to present the research as clearly and accurately as possible. I am grateful to Tehnobiro d.o.o., the Republic of Slovenia's Public Scholarship, Development, Disability, and Maintenance Fund, the Dutch Ministry of Foreign Affairs, and IHE Delft for funding the project that led to this dissertation.

I consider myself fortunate to have had the opportunity to work with the Tehnobiro d.o.o (Maribor, Slovenia) team, who taught me the ins and outs of engineering trade, from designing, construction, manufacturing, to testing of the system. Your contributions have increased the depth and breadth of this work, and most importantly provided me with tools that will help me in my future endeavors. Of course, the smooth operation of the system would not have been possible without the assistance of the Ptuj wastewater treatment plant team (Ptuj, Slovenia), Janko, Roman, Dani, Branko, Jože, Jernej, Janez and many others, who provided technical and laboratory support during my research field investigations in Ptuj, Slovenia. Thank you also to the team Department of Energy Engineering, German Jordanian University for your assistance during the testing phase of the system in Jordan Valley, Jordan.

I would like to express my gratitude to my family and friends for their unending support. Mau, Marko, Mohaned, Andreas, Megan, Nasir, Shwetha, Shekhar, Tudor, Ivan, and Inaki, thank you for listening, your support, love, shared smiles, laughs, advices, ideas, tears, meals and, for the countless stories we have created along the way. I am looking forward to our new adventures together!

At last, I would like to express my gratitude to my family. Father and brother, I am grateful for our brainstorming sessions. We discussed research-related topics ranging from experiment design to data interpretation. Thank you, mother, for being my emotional support pillar, always lending an understanding ear when I needed it. Thank you also to my lovely niece for providing occasional distractions from my work, filling my days with dancing and singing.

*Netherlands Research School for the
Socio-Economic and Natural Sciences of the Environment*

D I P L O M A

for specialised PhD training

The Netherlands research school for the
Socio-Economic and Natural Sciences of the Environment
(SENSE) declares that

Eva Kocbek

born on 4 June 1991 in Maribor, Slovenia

has successfully fulfilled all requirements of the
educational PhD programme of SENSE.

City, XX Month 2021

Chair of the SENSE board

Prof. dr. Martin Wassen

The SENSE Director

Prof. Philipp Pattberg

The SENSE Research School has been accredited by the Royal Netherlands Academy of Arts and Sciences (KNAW)

K O N I N K L I J K E   N E D E R L A N D S E
A K A D E M I E   V A N   W E T E N S C H A P P E N

The SENSE Research School declares that Eva Kocbek has successfully fulfilled all
requirements of the educational PhD programme of SENSE with a
work load of 39.1 EC, including the following activities:

SENSE PhD Courses

- Environmental research in context (2017)
- Research in context activity: 'A presentation video of a novel mobilized semi-decentralized system, designed for the treatment of faecal sludge from on-site sanitation facilities and municipal sludge from wastewater treatment facilities' (2019)

Other PhD and Advanced MSc Courses

- Problem solving and decision-making in research, TU Delft (2016)
- Research Design, TU Delft (2016)
- Speedreading and mindmapping, TU Delft (2016)
- Using Creativity to Maximize Productivity and Innovation in Your PhD, TU Delft (2016)
- How to become effective a network conversation, TU Delft (2017)
- The Art of presenting science, TU Delft (2017)
- Presenting: Storyline Structure and Presentation Slides, TU Delft (2017)
- Foundation of teaching, learning and assessment, TU Delft (2019)

External training at a foreign research institute

- Solidorks and AutoCAD Modelling (2017)
- Applied work activities at Tehnobiro d.o.o., Slovenia (2018-2019)
- Faecal sludge management & Design of Sanitation Systems and Technologies, Ecole Polytechnique Fédérale de Lausanne (2021)

Management and Didactic Skills Training

- Supervising MSc student with thesis entitled 'Ceramic membranes for direct ultrafiltration of septage: The case of the UF unit of the Shit Killer for filtration of condensate from the microwave unit and filtrate from the press filter' (2019)
- Teaching 'Theory and operation of dewatering unit, ultrafiltration and RO membranes, microwave unit, combustion unit', German Jordanian University, Jordan (2020)

Impact and outreach

- Designing and writing an operational manual Slovenia (2019)
- Co-writing a popular article on https://www.sludgeprocessing.com (2020)

Oral Presentations

- *Mobile approach for faecal sludge treatment using microwave irradiation.* H2O Summit International Water Congress. 18- 21 April 2018, Rovinj, Croatia

SENSE coordinator PhD education

Dr. ir. Peter J. Vermeulen